# Robot Rights

# Robot Rights

David J. Gunkel

The MIT Press
Cambridge, Massachusetts
London, England

© 2018 Massachusetts Institute of Technology

All rights reserved. No part of this book may be reproduced in any form by any electronic or mechanical means (including photocopying, recording, or information storage and retrieval) without permission in writing from the publisher.

This book was set in ITC Stone Serif Std by Toppan Best-set Premedia Limited. Printed and bound in the United States of America.

Library of Congress Cataloging-in-Publication Data

Names: Gunkel, David J., author.
Title: Robot rights / David J. Gunkel.
Description: Cambridge, MA : MIT Press, [2018] | Includes bibliographical references and index.
Identifiers: LCCN 2018013459 | ISBN 9780262038621 (hardcover : alk. paper)
Subjects: LCSH: Robots--Human factors. | Robots--Moral and ethical aspects.
Classification: LCC TJ211.49 .G88 2018 | DDC 179--dc23 LC record available at https://lccn.loc.gov/2018013459

ISBN: 978-0-262-03862-1

10 9 8 7 6 5 4 3 2 1

# Contents

Preface  ix
Acknowledgments  xiii

**Introduction  1**
**1  Thinking the Unthinkable  13**
   1.1  Robot  14
       1.1.1  Science Fiction  15
       1.1.2  Indeterminate Determinations  19
       1.1.3  Moving Target  22
       1.1.4  Results/Summary  23
   1.2  Rights  26
       1.2.1  Definition  26
       1.2.2  Theories of Rights  30
   1.3  Robot Rights or the Unthinkable  33
       1.3.1  Ridiculous Distractions  34
       1.3.2  Justifiable Exclusions  38
       1.3.3  Literal Marginalization  43
       1.3.4  Exceptions that Prove the Rule  44
   1.4  Summary  48
**2  !S1→!S2: Robots Cannot Have Rights; Robots Should Not Have Rights  53**
   2.1  Default Understanding  53
   2.2  Literally Instrumental  55
       2.2.1  Being vs. Appearance  57
       2.2.2  Ontology Precedes Ethics  59
       2.2.3  Limited Rights  60

2.3 Instrumentalism at Work  62
    2.3.1 Expertise  63
    2.3.2 Robots Are Tools  64
    2.3.3 Is/Ought Inference  65
2.4 Duty Now and for the Future  66
2.5 Complications, Difficulties, and Potential Problems  69
    2.5.1 Tool != Machine  69
    2.5.2 Not Just Tools  72
    2.5.3 Ethnocentrism  75
2.6 Summary  76

# 3 S1→S2: Robots Can Have Rights; Robots Should Have Rights  79

3.1 Evidence, Instances, and Examples  80
    3.1.1 Philosophical Arguments  80
    3.1.2 Legal Arguments  88
    3.1.3 Common Features and Advantages  92
3.2 Complications, Difficulties, and Potential Problems  93
    3.2.1 Infinite Deferral  94
    3.2.2 Is/Ought Inference  95
3.3 Summary  105

# 4 S1 !S2: Although Robots Can Have Rights, Robots Should Not Have Rights  107

4.1 The Argument  107
4.2 Complications, Difficulties, and Potential Problems  110
    4.2.1 Normative Proscriptions  111
    4.2.2 Ethnocentrism  115
    4.2.3 Slavery 2.0  117
4.3 Summary  130

# 5 !S1 S2: Even If Robots Cannot Have Rights, Robots Should Have Rights  133

5.1 Arguments and Evidence  134
    5.1.1 Anecdotes and Stories  134
    5.1.2 Scientific Studies  135
    5.1.3 Outcomes and Consequences  138
5.2 Complications, Difficulties, and Potential Problems  142
    5.2.1 Moral Sentimentalism  142
    5.2.2 Appearances  145

      5.2.3 Anthropocentrism, or "It's Really All About Us"   150
      5.2.4 Critical Problems   154
  5.3 Summary   157
**6 Thinking Otherwise   159**
  6.1 Levinas 101   160
      6.1.1 A Different Kind of Difference   161
      6.1.2 Social and Relational   165
      6.1.3 Radically Superficial   168
  6.2 Applied (Levinasian) Philosophy   170
      6.2.1 The Face of the Robot   171
      6.2.2 Ethics Beyond Rights   175
  6.3 Complications, Difficulties, and Potential Problems   177
      6.3.1 Anthropocentrism   177
      6.3.2 Relativism and Other Difficulties   180
  6.4 Summary   184

Notes   187
References   203
Index   235

# Preface

Whether we recognize it as such or not, we are in the midst of a robot invasion. The machines are now everywhere and doing virtually everything. We chat with them online, we play with them in digital games, we collaborate with them at work, and we rely on their capabilities to manage many aspects of our increasingly complex data-driven lives. Consequently, the "robot invasion" is not something that will transpire as we have imagined it in our science fiction, with a marauding army of evil-minded androids descending from the heavens. It is an already occurring event with machines of various configurations and capabilities coming to take up positions in our world through a slow but steady incursion. It looks less like *Battlestar Galactica* and more like the Fall of Rome.

Some statistics: Industrial Robots (IRs) have slowly but steadily been invading our work places since the mid-1970s, and this infiltration has, in recent years, appeared to have accelerated to impressive levels. As S. M. Solaiman (2017, 156) recently reported: "The International Federation for Robotics (IFR) in a 2015 report on IRs found an increase in the usage of robots by 29% in 2014, which recorded the highest sales of 229,261 units for a single year (IFR 2015). IFR estimates that about 1.3 million new IRs will be employed to work alongside humans in factories worldwide between 2015 and 2018 (IFR 2015). IFR has termed this remarkable increase as 'conquering the world' by robots (IFR 2015)." In addition to these industrial applications, there are also "service robots," which are characterized as machines involved in "entertaining and taking care of children and elderly people, preparing food and cooking in restaurants, cleaning residential premises, and milking cows" (Cookson 2015). There are, according to data provided by the Foundation for Responsible Robotics, 12 million service robot currently in operation across the globe, and the IFR predicts "an exponential rise" with the population of service robots expected to reaching 31 million units by 2018 (Solaiman 2017, 156).

We also are seeing a significant increase in the number and types of "social robots," like the PARO therapy robot, Jibo, "the world's first family robot," and the other tabletop assistants like Amazon's Echo/Alexa and Google Home. These devices are a subset of service robots specifically designed for human social interaction. And the predictions for these social robots exceed both IRs and service robots, with countries like South Korea aiming to put a robot in every home by 2020 (Lovgren 2006). Finally, there are distributed systems like the Internet of Things (IoT), where numerous connected devices "work together" to make an automated arrangement that is not "a robot" in the typical sense of the word but a network of interacting and smart devices. The Internet is already overrun, if not entirely run, by machines, with better than 50% of all online activity being machine generated and consumed (Zeifman 2017), and it is now estimated that IoT will support over 26 billion interactive and connected devices by 2020 (by way of comparison, the current human population of planet earth is estimated to be 7.4 billion) (Gartner 2013).

As these various mechanisms take up increasingly influential positions in contemporary culture—positions where they are not necessarily just tools or instruments of human action but a kind of interactive social entity in their own right—we will need to ask ourselves some rather interesting but difficult questions. At what point might a robot, algorithm, or other autonomous system be held accountable for the decisions it makes or the actions it initiates? When, if ever, would it make sense to say, "It's the robot's fault"? Conversely, when might a robot, an intelligent artifact, or other socially interactive mechanism be due some level of social standing or respect? When, in other words, would it no longer be considered nonsense to inquire about the rights of robots?—and to ask the question: "Can and should robots have rights?"

Although considerable effort has already been expended on the question of robots and responsibility, the other question, the question of *robot rights*, remains conspicuously absent or at least marginalized. In fact, for most people, "the notion of robots having rights is unthinkable," as David Levy (2005, 393) has asserted. This book is designed as both a provocation and thoughtful response to this apparent "fact." It is, therefore, deliberately designed to think the unthinkable by critically considering and making (or venturing to make) a serious philosophical case for the rights of robots. It does so not to be controversial, even if controversy is often the result of this kind of philosophical intervention, but in order to respond to some very real and pressing challenges concerning emerging technology and the current state of and future possibilities for moral reasoning.

## Acknowledgments

I have been at the task of "robot rights" for some time. Its first formal articulation occurred in the last chapter of the book *Thinking Otherwise: Philosophy, Communication, Technology* (Purdue University Press, 2007). That final chapter eventually spilled over into an entire book-length analysis, *The Machine Question: Critical Perspective on AI, Robots and Ethics* (MIT Press, 2012), which in turn generated a number of subsequent essays resulting from my efforts to respond to and take responsibility for important questions and criticisms that the book had made possible. These essays include: "A Vindication of the Rights of Machines" (Gunkel 2014a); "The Other Question: The Issue of Robot Rights" (Gunkel 2014b); "Apocalypse Not, or How I Learned to Stop Worrying and Love the Machine" (Gunkel and Cripe 2014); "The Rights of Machines: Caring for Robotic Care-Givers" (Gunkel 2015); and "Another Alterity: Rethinking Ethics in the Face of the Machine" (Gunkel 2016c).

This book is a continuation and culmination of that effort. Its current form and configuration are the direct result of a plenary address developed for and presented at *Robophilosophy/TRANSOR 2016: What Social Robots Can and Should Do?* This conference (the second in what is now a series of Robophilosophy conferences) was convened at the University of Aarhus, Denmark in October of 2016. My sincere thanks to Johanna Seibt and Marco Nørskov, who organized this event and in so doing created a unique venue for the kind of discussions and debates that have made work like this possible. A debt of gratitude is also due to the other presenters and participants at the conference who provided engaging opportunities for conversation throughout the event and responded to the presentation with thought-provoking questions and comments.

These critical rejoinders stayed with me after the conference and accompanied me on my way to the University of Vienna, where I had the good fortune to be able to brainstorm with Mark Coeckelbergh. Mark has been and continues to be a brilliant "sounding board" for bouncing around ideas, and it was due to these interactions in Vienna that it first became clear to me that this book needed to be the next writing project. In an effort to respond to Mark's insightful provocations and to test the basic concept for the book, I developed the conference presentation into a full paper shortly after returning home to Chicago. The paper (which bears the title "The Other Question: Can and Should Robots have Rights?") was completed in January of 2017, published in *Ethics and Information Technology* in the fall of that year (Gunkel 2017a), and thoughtfully reviewed by John Danaher (2017b) on his "Philosophical Disquisitions" blog.

While the paper—which is a sort of sketch of what is presented here in greater detail—was working its way through the submission process at the journal, I began formulating ideas for the book. This material was then shepherded through the proposal and review process by Philip Laughlin at the MIT Press. I have now worked with Philip on three titles, and I am, every single time, grateful for his knowledge and intuitive grasp of the strengths and weaknesses of a project and his ability to tap the best peer-reviewers. Like the two previous books, *Robot Rights* would not have been possible without Philip's insights and "behind the scenes" effort and guidance.

The writing of the manuscript took place during the spring and summer of 2017 and was supported by the generous gift of time provided by way of sabbatical leave from Northern Illinois University. The final version was ushered through the twists and turns of the production process by Marcy Ross, and the index was assembled by Danielle Watterson. Like all previous publication efforts, this entire enterprise would not have been possible without the love and supported of my family: Ann Hetzel Gunkel, Stanisław Gunkel, and Maki—the German shorthaired pointer, in whose face I have had to face all kinds of questions regarding who or what can and should be Other.

http://machinequestion.org/robotrights

# Introduction

The majority of work concerning the ethics of artificial intelligence and robots focuses on what philosophers call an agent-oriented problematic. This is true for what Gianmarco Veruggio (2005) called "roboethics"; for what Michael and Susan Leigh Anderson (2011) have designated *Machine Ethics*; for what Wendell Wallach and Colin Allen (2009, 4) propose with the concept "artificial moral agent" or AMA; and for *Robot Ethics* as developed by Patrick Lin et al. (2012 and 2017). It also holds for most of the work that is currently being done in robot law, policy, and ethics. There are, for instance, a growing number of publications concerning: the safety and reliability of self-driving vehicles (Hammond 2016, Lin 2016, Nyholm and Smids 2016, Casey 2017); efforts to engineer intelligent systems that are designed to value human life, or what is called "friendly AI" (Yudkowsky 2001, Rubin 2011, Muehlhauser and Bostrom 2014); a seemingly inexhaustible supply of predictions of technological unemployment and the potentially adverse consequences of increased automation for individuals and human society (Brynjolfsson and McAfee 2011, Ford 2015, Frey and Osborne 2017, LaGrandeur and Hughes 2017); and debates over the social costs and consequences of autonomous weapon systems (Arkin 2009, Sharkey 2012, Bhuta et al. 2016, Leveringhaus 2016, Krishnan 2016), child- and elder-care robots (Sparrow and Sparrow 2006, Anderson and Anderson 2008, Sharkey and Sharkey 2010, Beran and Ramirez-Serrano 2010, Bemelmans et al. 2010, Whitby 2010, Sharkey and Sharkey 2012), social and sociable robots (Breazeal 2002, Nørskov 2016, Seibt et al. 2016), and even sex machines (Levy 2007, Samani et al. 2010, Sullins 2012, Ess 2016, Richardson 2016a and 2016b, Lee 2017, Sharkey et al. 2017, Danaher and McArthur 2017). The organizing question behind many of these efforts is: "What can and should robots do?" This is principally an inquiry regarding the opportunities and challenges of machine action or agency, and it is important insofar as it concerns the impact and consequences of robots in our world.[1]

But this is only one half of the story. As Steve Torrance (2008, 506) and Luciano Floridi (2013, 135–136) remind us, moral situations involve at least two interacting components—the initiator of the action or the *agent* and the receiver of this action or the *patient*. So far, the vast majority of efforts in roboethics, robot ethics, machine ethics, military and social robotics, etc., have been primarily concerned with robots as agents, or what Torrance (2008, 505–506) calls the "producer" or "source" of moral/legal action. What I propose to do in this book is shift the focus and consider things from the other side—the side of machine patiency or what Torrance (2008, 505–506) calls, by contrast, "moral consumers" or the "target" of moral/legal action.[2] Doing so necessarily entails a related but entirely different set of variables and concerns. The operative question for a patient-oriented investigation is not "What robots can and should do?" but "How can and should we respond to these artifacts?" How can and should we make a response in the face of the robot, especially the increasingly sociable robot, that is, as Cynthia Breazeal (2004, 1) describes it, intentionally designed to be "socially intelligent in a human like way" such that "interacting with it is like interacting with another person." Or to put it in terms of a question: "Can and should robots have rights?"

## Research Question

This question—"Can and should robots have rights?"—consists of two separate queries: "Can robots have rights?" which is a question that asks about the ontological capabilities of a particular entity; and "Should robots have rights?" which is a question that inquiries about normative obligations in the face of this entity. These two questions invoke and operationalize a rather famous conceptual distinction in philosophy that is called the is/ought problem, or Hume's Guillotine. In *A Treatise of Human Nature* (first published in 1738), David Hume differentiated between two kinds of statements: descriptive statements of fact and normative statements of value (Schurz 1997, 1). For Hume, the problem was the fact that philosophers, especially moral philosophers, often fail to distinguish between these two kinds of statements and therefore slip imperceptibly from one to the other:

# Introduction

In every system of morality, which I have hitherto met with, I have always remarked, that the author proceeds for some time in the ordinary ways of reasoning, and establishes the being of a God, or makes observations concerning human affairs; when all of a sudden I am surprised to find, that instead of the usual copulations of propositions, *is*, and *is not*, I meet with no proposition that is not connected with an *ought*, or an *ought not*. This change is imperceptible; but is however, of the last consequence. For as this *ought*, or *ought not*, expresses some new relation or affirmation, 'tis necessary that it should be observed and explained; and at the same time that a reason should be given, for what seems altogether inconceivable, how this new relation can be a deduction from others, which are entirely different from it. But as authors do not commonly use this precaution, I shall presume to recommend it to the readers; and am persuaded, that this small attention would subvert all the vulgar systems of morality, and let us see, that the distinction of vice and virtue is not founded merely on the relations of objects, nor is perceived by reason (Hume 1980, 469).

In its original form, Hume's argument may appear to be rather abstract and indeterminate. But his point becomes immediately clear, when we consider an actual example, like the arguments surrounding the politically charged debate about abortion. "The crucial point of this debate," as Gerhard Schurz (1997, 2) explains, "is the question which factual property of the unborn child is sufficient" for attributing to it an unrestricted right to life.

For the one party in this debate, which often appeals to the importance of our moral conscience and instinct, it is obvious that this factual property is the *fertilization*, because at this moment a human being has been created and life begins. Going to the other extreme, there are philosophers like Peter Singer or Norbert Hoerster who have argued that this factual property is the beginning of the *personality* of the baby, which includes elementary self-interests as well as an elementary awareness of them. From the latter position it unavoidably follows that not only embryos but even very young babies, which have not developed these marks of personality, do not have the unrestricted right to live which older children of adults have (Schurz 1997, 2–3).

What Schurz endeavors to point out by way of this example is that the two sides of the abortion debate—the two different and opposed positions concerning the value of an unborn human fetus—proceed and are derived from different ontological commitments concerning what exact property or properties count as morally significant. For one side, it is the mere act of embryonic fertilization; for the other, it is the acquisition of personality and the traits of personhood. Consequently, ethical debate—especially when it concerns the rights of others—is typically, and Hume would say unfortunately, predicated on different ontological assumptions of fact that are then taken as the rational basis for moral value and decision-making.

Since Hume, there have been numerous attempts to resolve this "problem" by bridging the gap that is supposed to separate "is" from "ought" (cf. Searle 1964). Despite these efforts, however, the problem remains, and is considered one of those important and intractable philosophical dilemmas that people write books about (Schurz 1997 and Hudson 1969). As Schurz (1997, 4) characterizes it: "probably the most influential debate on the is-ought problem in our time is documented by Hudson (1969, reprint 1972, 1973, 1979). Here Black and Searle have tried to show that logically valid is-ought-inferences are indeed possible, whereas Hare, Thomson and Flew have defended Hume's thesis and tried to demonstrate that Black's and Searle's arguments are invalid."

For our purposes what is important is not the logical complications of the "is-ought fallacy" as Hume originally characterized it or the ongoing and seemingly irresolvable debate concerning the validity of the is-ought inference as it has been developed and argued in the subsequent literature. What is pertinent for the investigation at hand is to recognize how the verbs "is" and "ought" organize qualitatively different kinds of statements and modes of inquiry. The former concerns ontological matters or statements of fact; the latter consists in axiological decisions involving what should be done or what ought to be done. The guiding question of our inquiry utilizes modal variants of these two verbs, namely "can" and "should." Employing Hume's terminology, the question "Can robots have rights?" may be reformulated as: "Are robots capable of being holders of rights?" Or, more specifically: "Are robots capable of having one or more of the Hohfeldian incidents (see chapter 1), namely, privileges, claims, powers, and/or immunities?" And the question "Should robots have rights?" can be reformulated as "Ought robots be considered holders of rights?" Or, more specifically: "Ought robots be considered as having privileges, claims, powers, and/or immunities?" The modal verb "can," therefore, asks an ontologically oriented question about the factual capabilities or properties of the entity, while "should" organizes an inquiry about axiological issues having to do with obligations to this entity. Following the Humean thesis, therefore, it is possible to make the following distinction between two different kinds of statements:

S1 = "Robots can have rights."

S2 = "Robots should have rights."

Because the is-ought inference is and remains an undecided and open question (Schurz 1997, 4), it is possible to relate S1 to S2 in ways that generate four options or modalities concerning the moral situation of robots. These four modalities can be organized into two pairs. In the first pair, which upholds and supports the is-ought inference, the affirmation or negation of the ontological statement (S1) determines the affirmation or negation of the axiological statement (S2). This can be written (using a kind of pseudo object-oriented programming code) in the following way:

!S1→!S2—"Robots cannot have rights. Therefore robots should not have rights."

S1→S2—"Robots can have rights. Therefore robots should have rights."

In the second pair, which endorses the Humean thesis or contests the inference of ought from is, one affirms the ontological statement (S1) while denying the axiological statement (S2), or vice versa. These two modalities may be written in the following way:

S1 !S2—"Although robots can have rights, they should not have rights."

!S1 S2—"Even if robots cannot have rights, they should have rights."

**Organization and Method**

In the chapters that follow, we will critically evaluate each one of these modalities as they are deployed and developed in the literature, performing a kind of cost/benefit analysis of the available arguments concerning the rights (or lack thereof) of robots and similar forms of autonomous technology. Prior to engaging in this effort, however, the first chapter will pause to consider and deal with the two words that comprise the terms of the investigation. Both "robot" and "rights" are nouns with complex denotations, and their conjunction, as it occurs and is announced in the title to the book, is something that many theorists and practitioner find "unthinkable" for one reason or another. The first chapter, therefore, does not simply define terminology, but grapples with and accounts for the difficulties of language that are already part and parcel of this investigation, or any philosophical investigation, since at least the time of the linguistic turn (if not well in advance of this development by way of what was already articulated in Plato's *Cratylus*). Words matter, and the words "robot" and "rights" have complicated histories, definitions, and employments. The first chapter engages in this problematic and accounts for its necessary influence and consequences.

After getting a handle on the terminology, the subsequent chapters consider each one of the four modalities derived from the analysis of the Humean thesis. The second and third chapters consider those proposals and arguments where the derivation of *ought* from *is* has been upheld, operationalized, and/or defended. In particular, chapter 2 examines situations where a negative response to the ontological question—"Can robots have rights?"—produces or entails a negative response to the axiological question—"Should robots have rights?" Conversely, the third chapter examines situations where an affirmative response to the ontological question produces or entails an affirmative response to the axiological question: "Robots can have rights. Therefore robots should have rights." The second set of chapters—chapters four and five—consider those texts and arguments where the ontological and axiological dimensions are distinguished and held apart or in isolation from each other. Chapter 4 considers what results when one affirms that robots can in fact have rights, but simultaneously maintains that this does not necessitate nor imply that they should be extended any form of moral consideration or legal concern. Chapter 5 looks at the opposite situation where the ontological condition—"Robots cannot have rights."—does not limit or impede one from extending rights to robots and affirming the fact that robots (or at least some particular kinds of robots) should have rights.

In all four chapters, the objective is not to select sides by advocating for one or the other or trying to mediate their differences and disputes. The goal is to perform a cost/benefit analysis of each modality in an effort to understand the opportunities and challenges of each one and to map the terrain, or (if one would prefer to avoid reliance on this kind of cartographic metaphor) to identify the range of possibilities, of and for robot rights. In terms of its methodology, this effort is *philosophical* in the strict sense of the word. In fact, it is fundamentally Socratic. In Plato's *Apology*, Socrates offers his fellow citizens an account of his "Herculean labors" (22a). He narrates how his friend and colleague, the late Chaerephon, once asked the oracle at Delphi whether there was anyone wiser than the man Socrates. He explains how, much to his own surprise, the Delphic Pythia replied that there was no one wiser. And he describes how, since that time, he has been traveling throughout the city of Athens looking to prove the oracle incorrect by finding someone who was indeed wiser. This effort, as Socrates tells the story, leads him to seek out and dialogue with the most knowledgeable and learned individuals of his time: the politicians and rhetoricians, the scientists and teachers, the tragic poets and artists, and the craftsmen and engineers. But the result of these inquiries, as Socrates recounts, always left him wanting; that is, those who seemed best situated to know about important matters—things like truth, justice, knowledge, and beauty—inevitably betrayed some complication, deficiency, or problem in their own knowledge, terminology, or way of thinking.

Following this precedent, chapters 2–5 consist in a similar kind of critical endeavor, seeking out and consulting knowledgeable experts in order to gather up the best thinking about robots and rights as it is currently formulated by those individuals and groups of individuals who are best situated to know such things: philosophers, ethicists, jurists, engineers, scientists, etc. However, because we, unlike Socrates in the Platonic dialogues, live in a time and culture that is literate, this investigation will not engage in dialogue with individuals (in other words, what follows does not consist of personal conversations or interviews, which would be another way to do this kind of research) but with the various texts and documents—books, scientific journals, popular publications, films and television programs, radio transmissions and podcasts, blogs and websites, etc.—that have been produced on the subject of robot rights.[3] The product of this investigation is similar to what appears in the Platonic dialogues—the record of a detailed and probing interrogation that seeks out the best thinking that is currently available. But as was the case with the Socratic investigations, it will not be surprising if each modality leads, in the final analysis, to a kind of incomplete response that leaves us wanting.

Although each modality has its own particular set of advantages (and attendant limitations), none of the four will provide what would be considered a definitive case either for or against robot rights. In response to this, one could obviously continue to develop arguments and accumulate evidence supporting one modality or another. But the effect of this kind of effort, although entirely reasonable and justified, will be simply to elaborate what has already been formulated and therefore will not necessarily advance the conversation much further than what has already been undertaken and achieved. In order to get some new perspective on the issue, the book ends by trying something different. This alternative, which I will call *thinking otherwise*,[4] does not argue in favor of or in opposition to the is-ought inference but takes aim at this conceptual arrangement and apparatus. It does so by deliberately flipping the Humean script, considering not "how *ought* may be derived (or not) from *is*" but rather "how *is* is only able to be derived from *ought*."

This operation is, technically speaking, a *deconstruction*. I have now introduced, characterized, and even apologized for the term "deconstruction" in virtually every book that bears my name.[5] In order to avoid simply repeating this material—although, as Gilles Deleuze (1994) has argued, there is a mode of repetition that is not necessarily more of the same or the mere opposite of difference (cf. Gunkel 2016 for more on this subject)—we can (and should) briefly take note of four related items:

1. *Word Origin*. The word "deconstruction," as its progenitor, Jacques Derrida, has explained and acknowledged, was originally appropriated from and devised as means of translating Martin Heidegger's *Destruktion* and the task of critically analyzing the history of Western ontology, which was projected to have been the subject of the second volume of *Being and Time* ("projected" because this part of the endeavor was never actually completed or undertaken as such). "When I choose this word," as Derrida (1991, 271) once explained, "or when it imposed itself on me—I think it was in *Of Grammatology* —I little thought it would be credited with such a central role in the discourse that interested me at the time. Among other things I wished to translate and adapt to my own ends the Heideggerian word *Destruktion*."
2. *Negative Characterization*. Following from this, it can be said that "deconstruction" does not mean to take apart, to un-construct, or to disassemble. Despite this rather widespread and popular misconception, which has become something of an institutional (mal)practice, deconstruction is NOT a form of destructive analysis, a kind of demolition, or the process of reverse engineering. "The de- of deconstruction," as Derrida (1993, 147) has said quite explicitly (and on more than one occasion), "signifies not the demolition of what is constructing itself, but rather what remains to be thought beyond the constructionist or destructionist schema." Deconstruction, therefore, names something entirely other than what is commonly understood and delimited by

the conceptual opposition situated between, for example, "construction" and "destruction."
3. *Positive Characterization.* Deconstruction names a way—what Derrida (1982, 41) calls a "general strategy"—to intervene in "the binary oppositions of metaphysics," those conceptual pairings like construction/destruction, good/bad, self/other, is/ought, etc. Deconstruction, then, constitutes a mode of critical intervention that takes aim at these conceptual oppositions and does so in a way that does not simply neutralize them or remain within the hegemony of the existing order. In order to do this, deconstruction consists in a double gesture or what Derrida (1982, 41) also calls "a double science." This two-step procedure necessarily begins with a phase of inversion, where a particular duality or conceptual opposition is deliberately overturned by siding with the traditionally deprecated term. This is, quite literally, a revolutionary operation insofar as the existing order is deliberately inverted or turned around. But this is only half the story; it is just one aspect or phase. Such conceptual inversion, like all revolutionary gestures—whether social, political, philosophical, or artistic—actually does little or nothing to challenge the dominant system. In merely exchanging the relative positions occupied by the two opposed terms, inversion still maintains, albeit in an inverted or reversed form, the conceptual opposition in which and on which it operates. Simply flipping the script, as Derrida (1981, 41) concludes, still "resides within the closed field of these oppositions, thereby confirming it." For this reason, deconstruction also entails a second phase or operation. "We must," as Derrida (1981, 42) describes it, "also mark the interval between inversion, which brings low what was high, and the irruptive emergence of a new 'concept,' a concept that can no longer be, and never could be, included in the previous regime." This new "concept" is, strictly speaking, no concept whatsoever (which does not mean that it is simply the opposite of the conceptual order), for it always and already exceeds the system of dualities that define the conceptual order as well as the nonconceptual order with which the conceptual order has been articulated (Derrida 1982, 329). This "new concept" (that is strictly speaking not really a concept) is what Derrida (1981, 43) also calls an "undecidable." It is, first and foremost, that "that can no longer be included within philosophical (binary) opposition, but which, however, inhabits

philosophical opposition, resisting and disorganizing it, without ever constituting a third term, without ever leaving room for a solution in the form of speculative dialectics." It is, in other words, a way of "thinking outside the box, where 'the box' is defined as the total enclosure that delimits the very possibility of any kind of thought whatsoever" (Gunkel 2001, 203).

4. *Example.* Perhaps the best example or illustration of deconstruction's two-step operation is available with the term "deconstruction" itself. In a first move, deconstruction flips the script by putting emphasis on the negative term "destruction" as opposed to construction. In fact, the apparent similitude between the two words "deconstruction" and "destruction" is a deliberate and calculated aspect of this effort. But this is only step one. In the second move of this "double science," deconstruction introduces a brand new concept, the concept of "deconstruction." The novelty of this concept is marked, quite literally, in the material of the word itself. "Deconstruction," which is fabricated by combining the "de-" of "destruction" and attaching it to the opposite term "construction," is a neologism that does not quite fit in the existing order of things. It is an exorbitant and intentionally undecidable third term that names a new kind of concept. This new concept, despite first appearances, is not the mere opposite of construction but exceeds the conceptual order instituted and regulated by the terminological opposition situated between construction/destruction.

In both reversing and displacing the Humean thesis—privileging *ought* and making it determinative of *is*—ethics precede ontology. *What the other is* is something that is first and foremost decided on the basis of *how it is treated* in actual social situations and circumstances and dependent upon how we have decided to respond in the face of the *Other*. This alternative, this version of a "social relational" ethic—which will be (and cannot help but be) disorienting for standard modes of moral reasoning—will be pursued and developed by way of an engagement with the writings of Emmanuel Levinas and those others who follow his lead. And this will be pursued and accomplished even if doing so will necessitate a certain unorthodox reading of Levinas's texts and a critical break with essential (but exceedingly problematic) aspects of Levinasian philosophy.

## Outcomes and Results

In the end, what this effort produces is both more and less than what might have been anticipated at the beginning. It will be *less*, if what one expects from the book is either a simple "yes"/"no" response to the question "Can and should robots have rights?" or (what might be a more accurate and modest expectation) a set of normative rules, ethical codes of conduct, or even guidelines and frameworks for devising policy. As a book of philosophy, the objective of the investigation is not—at least not principally—to issue normative prescriptions and administrative guidelines, but to unearth, examine, and evaluate the underlying conditions of possibility that already organize and structure (or even "disorganize" and "complicate") such efforts. Consequently, and similar to what was pursued in *The Machine Question* (Gunkel 2012), this book is *critical* in the strict philosophical sense of the word. As Barbara Johnson (1981, xv) accurately characterizes it, a *critique* is not simply the examination of a particular system's flaws and imperfections, designed to make things better. It is not, in other words, an effort at rudimentary fault-finding or remedial debugging. Instead, "it is an analysis that focuses on the grounds of that system's possibility. The critique reads backwards from what seems natural, obvious, self-evident, or universal, in order to show that these things have their history, their reasons for being the way they are, their effects on what follows from them, and that the starting point is not a given but a construct, usually blind to itself" (Barbara Johnson 1981, xv).

A critical approach to the question of robot rights, therefore, "reads backwards" from the assertions, arguments, and debates about robots and rights in order to demonstrate how many of the concepts, approaches, and even vocabularies that are considered to be natural, obvious, self-evident, and/or universal have their own complicated legacies and logics. These constructs, which already frame and control the debate, remain largely invisible (or just barely visible) by operating behind the scenes or in/at the margins. The critical task, therefore, is to identify, explicate, and evaluate the oftentimes implicit operating systems that makes the discussion and debate about robots and rights possible in the first place. In other words (and in a modest effort to perform some "expectation management"), if what one seeks and hopes for in this book is an up/down vote on robot rights or a practical framework for devising and writing policy, then you will be disappointed.

But if what is sought is a deeper and more complete understanding of the opportunities and challenges made available by the questions concerning robot rights, then the book will deliver.

For this reason, the results of the analysis may actually be *more* than what is expected, insofar as this outcome makes a difference not just for the specific technology of robots but for moral philosophy in general. This means that there will be, following John Searle's (1980) original distinction regarding the science of AI, a *strong* and *weak* aspect to robot rights.[6] Efforts at what would be the "strong" variety involve those discussions and debates that seek to formulate policy, moral rules, and/or legal statutes regarding the social position and status of robots. And many of the texts that are considered and evaluated in what follows engage in this kind of undertaking. But these efforts also have a "weak" aspect ("weak" in Searle's particular sense of the word, not in the usual colloquial sense of being of lesser significance or importance) insofar as they put into play, operationalize, and stress test various conceptualizations of moral and legal rights, leading to critical and discerning reevaluations of how we have characterized these concepts in our own thinking and conversations. In other words, developing and debating the rights of robots does not necessarily take anything away from human beings and what (presumably) makes us special; it offers a critical tool for doing work in moral theory, making available new opportunities for us to be more precise and scientific about these distinguishing characteristics and their limits.

# 1 Thinking the Unthinkable

Before we get too far into investigating or asking about robot rights, it would be both expedient and prudent to clarify (or at least reflect critically on) what is meant by the seemingly simple words that comprise the terms of the investigation. "Robot" and "rights" are, at least at this point in time, both common and arguably well understood. We all know—or at least think we know—what a robot is. Science fiction literature, film, and digital media have been involved with imagining and imaging robots for close to a century. In fact, many of the most widely recognized and identifiable characters in twentieth and twenty-first century popular culture are robots: Robby the Robot of *Forbidden Planet*, Rosie from the *Jetsons*, *Tetsuwan Atomu* [Mighty Atom] or Astro Boy from Osamu Tezuka's long running manga series of the same name, William Joyce's *Rolie Polie Olie*, Lt. Commander Data of *Star Trek: The Next Generation*, R2-D2 and C-3PO of *Star Wars* ... and the list could go on.

"Rights" also appears to be a well-established and widely understood concept but for reasons that are rooted in the struggles of social reality rather than fiction. The history of the nineteenth and twentieth centuries can, in fact, be characterized as a sequence of conflicts over rights, or what Peter Singer (1989, 148) has called a "liberation movement," as previously excluded individuals and communities fought for and eventually achieved equal (or closer to equal) status, i.e., women, people of color, LGBTQ individuals, etc. No one it seems is confused by or unclear about the struggles for civil liberties and the criticisms of human rights abuses that have been, and unfortunately still are, part and parcel of everyday social reality.

Given prior experience with both terms, it would seem that there is not much with which to be concerned. But it is precisely this seemingly confident and unproblematic familiarity—this always-already way of understanding things—that is the problem. As Martin Heidegger (1962, 70) pointed out in the opening salvo of *Being and Time*, it is precisely those things that are closest to us and apparently well-understood in our everyday activities and operations that are the most difficult to see and explicitly conceptualize. For this reason, it is a good idea to begin by getting some critical distance on both terms in order to characterize explicitly what each one means individually and what their concatenation indicates for the effort that follows. Fortunately this "distancing" is already available and in play insofar as the phrase "robot rights" just "feels wrong" or is at least curious and thought-provoking. It is from this initial feeling of disorientation or strangeness—whereby what had been seemingly clear and unremarkable comes to stand out as something worth asking about—that one can begin to question and formulate what is meant by the terms "robot" and "rights."

## 1.1 Robot

"When you hear the word 'robot,'" Matt Simon (2017, 1) of *Wired* writes, "the first thing that probably comes to mind is a silvery humanoid, à la *The Day the Earth Stood Still* or C-3PO (more golden, I guess, but still metallic). But there's also the Roomba, and autonomous drones, and technically also self-driving cars. A robot can be a lot of things these days—and this is just the beginning of their proliferation. With so many different kinds of robots, how do you define what one is?" Answering this question is no easy task. In his introductory book on the subject, John Jordan (2016, 3–4) openly admits that there is already considerable indeterminacy and slippage with the terminology such that "computer scientists cannot come to anything resembling consensus on what constitutes a robot." The term "robot," then, is a rather noisy moniker with indeterminate and flexible semantic boundaries. And this difficulty, as Simon (2017, 1) points out, "is not a trivial semantic conundrum: Thinking about what a robot really is has implications for how humanity deals with the unfolding robo-revolution."

Rather than offering a definition, Jordan tries to account for and make sense of this terminological difficulty. According to his reading, "robot" is complicated for three reasons:

Reason 1 why robots are hard to talk about: the definition is unsettled, even among those most expert in the field.

Reason 2: definitions evolve unevenly and jerkily, over time as social context and technical capabilities change.

Reason 3: science fiction set the boundaries of the conceptual playing field before the engineers did. (Jordan 2016, 4–5)

These three aspects make defining and characterizing the word "robot" complicated but also interesting. We can, therefore, get a better handle on the problem of identifying what is and what is not a robot by considering each reason individually.

### 1.1.1 Science Fiction

Science fiction not only defines "the boundaries of the conceptual playing field," but is the original source of the term. The word *robot* came into the world by way of Karel Čapek's 1920 stage play, *R.U.R.* or *Rossumovi Univerzální Roboti* (*Rossum's Universal Robots*) in order to name a class of artificial servants or laborers. In Czech, as in several other Slavic languages, the word *robota* (or some variation thereof) denotes "servitude or forced labor." Čapek, however, was not (at least according to his own account) the originator of the term. That honor apparently belongs to the writer's older brother, the painter Josef Čapek. Thirteen years after the publication of the play, which had been wildly successful at the time, Karel Čapek explained the true origin of the term in the pages of *Lidove Noviny*, a newspaper published in Prague:

> The author of the play *R.U.R.* did not, in fact, invent that word; he merely ushered it into existence. It was like this: the idea for the play came to said author in a single, unguarded moment. And while it was still warm he rushed immediately to his brother Josef, the painter, who was standing before an easel and painting away at a canvas till it rustled. "Listen, Josef," the author began, "I think I have an idea for a play." "What kind," the painter mumbled (he really did mumble, because at the moment he was holding a brush in his mouth). The author told him as briefly as he could. "Then write it," the painter remarked, without taking the brush from his mouth or halting work on the canvas. The indifference was quite insulting. "But," the author said, "I don't know what to call these artificial workers. I could call them *Labori*, but that strikes me as a bit bookish." "Then call them Robots," the painter muttered, brush in mouth, and went on painting. And that's how it was. Thus the word Robot was born; let this acknowledge its true creator. (Čapek 1935; also quoted in Jones 2016, 53)

Since the publication of Čapek's play, robots have infiltrated the space of fiction. But what exactly constitutes a robot differs and admits of a wide variety of forms, functions, and configurations. Čapek's robots, for instance, were artificially produced biological creatures that were humanlike in both material and form. This configuration persists with the bioengineered replicants of *Bladerunner* and *Bladerunnner 2049* (the film adaptations of Philip K. Dick's *Do Androids Dream of Electric Sheep?*) and the skin-job Cylons of *Battlestar Galactica*. Other fictional robots, like the chrome-plated android in Fritz Lang's *Metropolis* and C-3PO of *Star Wars*, as well as the 3D printed "hosts" of HBO's *Westworld* and the synths of Channel 4/AMC's *Humans*, are humanlike in form but composed of non-biological materials. Others that are composed of similar synthetic materials have a particularly imposing profile, like *Forbidden Planet's* Robby the Robot, Gort from *The Day the Earth Stood Still*, or the Robot from the television series *Lost in Space*. Still others are not humanoid at all but emulate animals or other kinds of objects, like the trashcan R2-D2, the industrial tanklike Wall-E, or the electric sheep of Dick's novella. Finally there are entities without bodies, like the HAL 9000 computer in *2001: A Space Odyssey*, with virtual bodies, like the Agents in *The Matrix*, or with entirely different kinds of embodiment, like swarms of nanobots.

In whatever form they have appeared, science fiction already—and well in advance of actual engineering practice—has established expectations for what a robot is or can be. Even before engineers have sought to develop working prototypes, writers, artists, and filmmakers have imagined what robots do or can do, what configurations they might take, and what problems they could produce for human individuals and communities. Jordan (2016, 5) expresses it quite well: "No technology has ever been so widely described and explored before its commercial introduction. …Thus the technologies of mass media helped create public conceptions of and expectations for a whole body of compu-mechanical innovation that *had not happened yet*: complex, pervasive attitudes and expectations predated the invention of viable products" (emphasis in the original). A similar point has been made by Neil M. Richards and William D. Smart (2016, 5), who approach the question from the legal side of things: "So what is a robot? For the vast majority of the general public (and we include most legal scholars in this category), we claim that the answer to this question is inescapably informed by what they see in movies, the popular media, and, to a lesser extent, in literature. Few people have seen an actual robot, so they must draw conclusions from the depictions of robots that they have seen.

Anecdotally, we have found that when asked what a robot is, people will generally make reference to an example from a movie: Wall-E, R2-D2, and C-3PO are popular choices."

Because of this, science fiction has been both a useful tool and a potential liability. Engineers and developers, for instance, often endeavor to realize what has been imaginatively prototyped in fiction. Take for example the following explanation provided by the roboticists Minoru Asada, Karl F. MacDorman, Hiroshi Ishiguro, and Yasuo Kuniyoshi (2001, 185):

> Robot heroes and heroines in science fiction movies and cartoons like *Star Wars* in US and *Astro Boy* in Japan have attracted us so much which, as a result, has motivated many robotic researchers. These robots, unlike special purpose machines, are able to communicate with us and perform a variety of complex tasks in the real world. What do the present day robots lack that prevents them from realizing these abilities? We advocate a need for cognitive developmental robotics (CDR), which aims to understand the cognitive developmental processes that an intelligent robot would require and how to realize them in a physical entity.

This "science fiction prototyping," as Brian David Johnson (2011) calls it, is rather widespread in the discipline even if it is not always explicitly called-out and recognized as such. As Bryan Adams, Cynthia Breazeal, Rodney Brooks, and Brian Scassellati (2000, 25) point out: "While scientific research usually takes credit as the inspiration for science fiction, in the case of AI and robotics, it is possible that fiction led the way for science."

In addition to influencing research and development programs, science fiction has also proven to be a remarkably expedient mechanism—perhaps even the preferred mechanism—for examining the social opportunities and challenges of technological innovation in AI and robotics. As Sam N. Lehman-Wilzig (1981, 444) once argued: "Since it is beyond human capability to distinguish a priori the truly impossible from the merely fantastic, all possibilities must be taken into account. Thus science fiction's utility in outlining the problem." And a good number of serious research efforts (in philosophy and law, in particular) have found it useful to call upon and employ existing narratives, like the robot stories of Isaac Asimov (i.e., Gips 1991, Anderson 2008, Haddadin 2014), the TV series *Star Trek* (i.e., Introna 2010, Wallach and Allen 2009, Schwitzgebel and Garza 2015), the film *2001: A Space Odyssey* (i.e., Dennett 1997, Stork 1997, Isaac and Bridewell 2017), and the reimagined *Battlestar Galactica* (Dix 2008, Neuhäuser 2015, Guizzo and Ackerman 2016), or even to fabricate their own fictional anecdotes and "thought experiments" (i.e., Bostrom 2014, Ashrafian 2015a and 2015b) as a way of introducing, characterizing, and/or investigating a problem, or what could, in the near term, become a problem. Here too, as Peter Asaro and Wendell Wallach argue, science fiction seems to have a taken a leading role:

> [The] philosophical tradition has a long history of considering hypothetical and even magical situations (such as the Ring of Gyges in Plato's *Republic*, which bestows invisibility on its wearer). The nineteenth and twentieth centuries saw many of the magical powers of myth and literature brought into technological reality. Alongside the philosophical literature emerged literatures of horror, science fiction, and fantasy, which also explored many of the social, ethical, and philosophical questions raised by the new-found powers of technological capabilities. For most of the twentieth century, the examination of the ethical and moral implications of artificial intelligence and robotics was limited to the work of science fiction and cyberpunk writers, such as Isaac Asimov, Arthur C. Clarke, Bruce Sterling, William Gibson, and Philip K. Dick (to name only a few). It is only in the late twentieth century that we begin to see academic philosophers taking up these questions in a scholarly way. (Asaro and Wallach 2016, 4–5)

Despite its utility, however, for many laboring in the fields of robotics, AI, human robot interaction (HRI), behavioral science, etc., the incursion of "entertainment" into the realm of the serious work of science is also a potential problem and something that must be, if not actively counteracted, then at least carefully bracketed and held in check. Science fiction, it is argued, often produces unrealistic expectations for and irrational fears about robots that are not grounded in or informed by actual science (Bartneck 2004, Kriz et al. 2010, Bruckenberger et al. 2013, Sandoval et al. 2014). "Real robotics," as Alan Winfield (2011a, 32) explains, "is a science born out of fiction. For roboticists this is both a blessing and a curse. It is a blessing because science fiction provides inspiration, motivation and thought experiments; a curse because most people's expectations of robots owe much more to fiction than reality. And because the reality is so prosaic, we roboticists often find ourselves having to address the question of why robotics has failed to deliver when it hasn't, especially since it is being judged against expectations drawn from fiction."[1] For this reason, science fiction is both a useful tool for and a significant obstacle to understanding what the term "robot" designates.

### 1.1.2 Indeterminate Determinations

Even when one consults knowledgeable experts, there is little agreement when it comes to defining, characterizing, or even identifying what is (or what is not) a robot. Illah Nourbakhsh (2013, xiv) writes: "Never ask a roboticist what a robot is. The answer changes too quickly. By the time researchers finish their most recent debate on what is and what isn't a robot, the frontier moves on as whole new interaction technologies are born." Indicative of this problem is a podcast from John Siracusa and Jason Snell called "Robot or Not?" The first episode explained the *raison d'être* of the program: "You and I battling through time as we fall deeper and deeper into a crevasse about whether something is a robot or not." In each episode, Siracusa and Snell focus on an object, either fictional, like the character Mr. Roboto from the 1983 Styx album *Kilroy Was Here* (episode 1), Darth Vader (episode 3), and the Daleks and Cybermen from *Dr. Who* (episode 43); or actual, like the Roomba (episode 8), Siri (episode 16), and Chatbots (episode 49); and then debate whether the object in question is a robot or not. This goes on for one hundred episodes over the course of two years. What is remarkable in this ongoing deliberation is not what determinations come to be made (i.e., Siri is not a robot, but a Roomba is a robot) or what criteria are deployed to make the determination (usually something having to do with independent vs. human-dependent control) but the fact that such an ongoing discussion is possible in the first place and that what is and what is not a robot requires this kind of discursive effort.

Dictionaries provide serviceable but arguably insufficient characterizations. The *Oxford English Dictionary* (2017), for example, defines "robots" in the following way:

1. A machine capable of carrying out a complex series of actions automatically, especially one programmable by a computer.

    1.1: (Especially in science fiction) a machine resembling a human being and able to replicate certain human movements and functions automatically.

    1.2: A person who behaves in a mechanical or unemotional manner.

The *Merriam-Webster Dictionary* (2017) offers a similar characterization:

1. a: A machine that looks like a human being and performs various complex acts (such as walking or talking) of a human being; also: a similar

but fictional machine whose lack of capacity for human emotions is often emphasized.

b: An efficient, insensitive person who functions automatically.

2. A device that automatically performs complicated, often repetitive tasks.
3. A mechanism guided by automatic controls.

These definitions are generally considered to be both too broad, insofar as they could be applied to any computer program, and too narrow, because they tend to privilege humanlike forms and configurations that are, beyond what is portrayed in science fiction, more the exception than the rule. In response to this, professional organizations and handbooks and textbooks in robotics offer up what is purported to be a more precise characterization. The International Organization for Standardization (ISO) provides the following definition for "robots and robotic devices" in ISO 8373 (2012): "An automatically controlled, reprogrammable, multipurpose, manipulator programmable in three or more axes, which may be either fixed in place or mobile for use in industrial automation applications." But this characterization can be faulted for being too specific and restrictive, applying only to industrial robots and therefore unable to accommodate other kinds of applications, like social and companion robots.

One widely cited source of a more general and comprehensive definition comes from George Bekey's *Autonomous Robots: From Biological Inspiration to Implementation and Control*: "In this book we define a robot as a machine that senses, thinks, and acts. Thus, a robot must have sensors, processing ability that emulates some aspects of cognition, and actuators" (Bekey 2005, 2). This "sense, think, act" or "sense, plan, act" (Arkin 1998, 130) paradigm has considerable traction in the literature—evidenced by the very fact that it constitutes and is called a *paradigm*:

> Robots are machines that are built upon what researchers call the "sense-think-act" paradigm. That is, they are man-made devices with three key components: "sensors" that monitor the environment and detect changes in it, "processors" or "artificial intelligence" that decides how to respond, and "effectors" that act upon the environment in a manner that reflects the decisions, creating some sort of change in the world around a robot. (Singer and Sagan 2009, 67)

When defining a robot, the "sense-think-act paradigm" is as close to consensus as one might find in terms of a robot differentiated from a computer system. (Wynsberghe 2016, 40)

There is no concise, uncontested definition of what a "robot" is. They may best be understood through the sense-think-act paradigm, which distinguishes robots as any technology that gathers data about the environment through one or more sensors, processes the information in a relatively autonomous fashion, and acts on the physical world. (Jones and Millar 2017, 598)

This definition is, as Bekey (2005, 2) recognizes, "very broad," encompassing a wide range of different kinds of technologies, artifacts, and devices. But it could be too broad insofar as it may be applied to all kinds of artifacts that exceed the proper limits of what many consider to be a robot. As John Jordan (2016, 37) notes: "the sense-think-act paradigm proves to be problematic for industrial robots: some observers contend that a robot needs to be able to move; otherwise, the Watson computer might qualify." The Nest thermostat provides another complicated case. "The Nest senses: movements, temperature, humidity, and light. It reasons: if there's no activity, nobody is home to need air conditioning. It acts: given the right sensor input, it autonomously shuts the furnace down. Fulfilling as it does the three conditions, is the Nest, therefore, a robot?" (Jordan 2016, 37). And what about smartphones? According to Joanna Bryson and Alan Winfield (2017, 117) these devices could also be considered robots under this particular characterization. "Robots are artifacts that sense and act in the physical world in real time. By this definition, a smartphone is a (domestic) robot. It has not only microphones but also a variety of proprioceptive sensors that let it know when its orientation is changing or when it is falling."

In order to further refine the definition and delimit with greater precision what is and what is not a robot, Winfield (2012, 8) offers the following list of qualifying characteristics:

A robot is:

1. An artificial device that can *sense* its environment and *purposefully* act on or in that environment;
2. An *embodied* artificial intelligence; or
3. A machine that can *autonomously* carry out useful work.

Although ostensibly another iteration of sense-think-act, Winfield adds an important qualification to his list—"embodiment"—making it clear that a software bot, an algorithm, or an AI implementation like Watson or AlphaGo are not robots, strictly speaking. This is by no means an exhaustive list of all the different ways in which "robot" has been defined, explained, or characterized. What is clear from this sample, however, is that the term "robot" is open to a considerable range of diverse and even different denotations. And these "definitions are," as Jordan (2016, 4) writes, "unsettled, even among those most expert in the field."

### 1.1.3 Moving Target

To further complicate matters, words and their definitions are not stable; they evolve over time, often in ways that cannot be anticipated or controlled. This means that the word "robot," like any word in any language, has been and will continue to be something of a moving target. What had been called "robot" at the time that Čapek introduced the word to the world is significantly different from what was called "robot" during the development of industrial automation in the latter decades of the twentieth century, and this is noticeably different from what is properly identified as "robot" at this particular point in time. These differences are not just a product of technological innovation; they are also a result of human perception and expectation. In an effort to illustrate this point, Jordan (2016, 4) offers the following observation from Bernard Roth of the Stanford Artificial Intelligence Laboratory (SAIL): "My view is that the notion of a robot has to do with which activities are, at a given time, associated with people and which are associated with machines." As relative capabilities evolve, so do conceptions. "If a machine suddenly becomes able to do what we normally associate with people, the machine can be upgraded in classification and classified as a robot. After a while, people get used to the activity being done by machines, and the device gets downgraded from 'robot' to 'machine'" (Jordan 2016, 4, quoting Bernard Roth). Consequently, what is and what is not a "robot" changes. The same object with the same operational capabilities can at one time be classified as a robot and at another time be regarded as something that is a mere machine (which is, on Roth's account, presumably "less than" what is called "robot"). These alterations in the status of the object are not simply a product of innovation in technological capability; they are also a consequence of social context and what people think about the object.

Brian Duffy and Gina Joue (2004, 2), make a similar point concerning machine intelligence. "Once problems, originally thought of as hard issues, are solved or even just understood better, they lose the 'intelligent' status and become simply software algorithms." At one point in time, for instance, playing championship chess was considered a sign of "true intelligence," so much so that experts in AI and robotics stated, quite confidently, that computers will never achieve (or at least were many decades away from achieving) championship-level play in the game of chess (Dreyfus 1992, Hofstadter 1979). Then, in 1997, IBM's DeepBlue defeated Gary Kasparov. In the wake of this achievement, concepts regarding intelligence and the game of chess have been recalibrated. DeepBlue, in other words, did not prove that the machine was as intelligent as a human being by way of playing a game previously thought to have been a sign of true intelligence. Instead it demonstrated, as Douglas Hofstadter (2001, 35) points out, "that world-class chess-playing ability can indeed be achieved by brute force techniques—techniques that in no way attempt to replicate or emulate what goes on in the head of a chess grandmaster." DeepBlue, therefore, transformed the hard problem of chess into just another clever computer application. What changed was not just the technological capabilities of the mechanism (and DeepBlue did, in fact, integrate a number of recent technical innovations), but also the way people thought about the game of chess and intelligence.

"Robot," therefore, is not some rigorously defined, singular kind of thing that exists in a vacuum. What is called "robot" is something that is socially negotiated such that word usage and terminological definitions shift along with expectations for, experience with, and use of the technology. Consequently, one needs to be sensitive to the fact that whatever comes to be called "robot" is always socially situated. Its context (or contexts, because they are always plural and multifaceted) is as important as its technical components and characterizations. What is and what is not a robot, then, is as much a product of science and engineering practice as it is an effect of changing social processes and interactions.

### 1.1.4 Results/Summary

"Robot" is complicated. It is something that is as much an invention of fiction as it is of actual R&D efforts. Its definition and characterization, even among experts in the field, is unsettled and open to considerable slippage and disagreements. And, like all terms, the way the word is used and that to which it applies changes with alterations in both the technology and social context. One typical and expedient response to this kind of terminological trouble, especially in a text or published study, is to begin by clarifying the vocabulary and/or offering an operational definition. George

Bekey's *Autonomous Robots: From Biological Inspiration to Implementation and Control* (2015, 2) provides a good example of how this operates: "In this book we define a robot as ..." This is something of a standard procedure, and different versions of this kind of terminological declaration are evident throughout the literature. Raya Jones (2016, 5), for instance, begins the book *Personhood and Social Robotics* with the following: "For the book's purpose, a serviceable definition of social robot is *physically embodied intelligent systems that enter social spaces in community and domestic settings*" (italics in the original). Likewise for Ryan Calo, A. Michael Froomkin, and Ian Kerr's edited collection of essays, *Robot Law* (2016), which operationalizes a version of the sense-think-act paradigm:

> Most people, and undoubtedly all the contributors to this volume, would agree that a man-made object capable of responding to external stimuli and acting on the world without requiring direct—some might say constant—human control was a robot, although some might well argue for a considerably broader definition. Three key elements of this relatively narrow, likely under-inclusive, working definition are: (1) some sort of sensor or input mechanism, without which there can be no stimulus to react to; (2) some controlling algorithm or other system that will govern the responses to the sensed data; and (3) some ability to respond in a way that affects or at least is noticeable by the world outside the robot itself. (Calo et al. 2016, xi)

These declarations and working definitions are both necessary and expedient. They are needed to get the analysis underway and to identify, if only for the purpose of the investigation at hand, the particular kinds of objects to be included in the examination and those other kinds of things that can be excluded from further consideration. In defining robots as "physically embodied intelligent systems," for example, Jones makes a decision that eliminates an entire class of objects from further consideration: "this excludes disembodied automated response systems, search engines, etc." (Jones 2015, 5–6). Such decisions are, as Jones recognizes, unavoidably exclusive. They make a decisive cut—*de + caedere* ("to cut")—that says, in effect, these entities are to be included in the consideration that follows, while these other seemingly related objects are to be excluded and not given further attention, at least not in this particular context. Despite its usefulness, however, this kind of decision-making does have problems and consequences. It is, whether we recognize it as such or not, an expression and exercise of power. In deciding what is included and what is excluded, someone, or some group, bestows on themselves the right to declare what is inside and what is to be eliminated and left outside of the examination. And these exclusive and exclusionary decisions matter, especially in the context of an investigation that seeks to be attentive to moral questions and the social/political complications involved in the exercise of power.

Instead of instituting this kind of univocal decision—even a temporary determination that is in operation for the time being—we can alternatively hold the term open and recognize the opportunity that is available with such plurivocity and semantic drift. "Robot" already allows for and encompasses a wide range of different concepts, entities, and characterizations. It therefore is already the site of a conversation and debate about technology and its social position and status, and we should not be too quick to close off the possibilities that this lexical diversity enables and makes available. As Andrea Bertolini (2013, 216) argues, "all attempts at providing an encompassing definition are a fruitless exercise: robotic applications are extremely diverse and more insight is gained by keeping them separate." For this reason, Bertolini (2013, 219) does not define robot once and for all but provides a classificatory schema that allows for and tolerates polysemia:

If, then, a notion of robot is to be elaborated for merely descriptive—thus neither qualifying nor discriminating—purposes, it may be as follows: a machine which (i) may either have a tangible physical body, allowing it to interact with the external world, or rather have an intangible nature—such as a software or program, (ii) which in its functioning is alternatively directly controlled or simply supervised by a human being, or may even act autonomously in order to (iii) perform tasks, which present different degrees of complexity (repetitive or not) and may entail the adoption of non-predetermined choices among possible alternatives, yet aimed at attaining a result or providing information for further judgment, as so determined by its user, creator or programmer, (iv) including but not limited to the modification of the external environment, and which in so doing may (v) interact and cooperate with humans in various forms and degrees.

Adopting this kind of descriptive (although much more complex) framework—one that is tolerant of and can accommodate a range or an array of different connotations and aspects—allows for a term, like "robot," to be more flexible and for the analysis that follows to be more responsive to the ways the word is actually utilized and applied across different texts, social contexts, research efforts, historical epochs, etc. To put it in the terminology of computer programming, it means that "robot" is not a *scalar*, a variable taking one value that can be declared and assigned right at the beginning. It is more like an *array*, a multi-value variable that permits and allows for a range of different but related characteristics.

```
var robot = new Array("sense-think-act," "embodied,"
"autonomous," …);
```

This does not mean, however, that anything goes and that "robot" is whatever one wants or declares it to be. It means, rather, paying attention to how the term "robot" comes to be deployed, defined, and characterized in the scholarly, technical, and popular literature, including fiction; how the term's connotations shift over time, across different contexts, and even (at times) within the same text; and how these variations relate to and have an impact on the options, arguments, and debates concerning the moral situation and status of our technological artifacts.

## 1.2 Rights

Like the term "robot," the word "rights" appears to identify something that is immediately familiar and seemingly well-understood. Conversations about rights and their recognition, protection, and/or violation is rather common in contemporary discussions about moral, legal, and political matters. "The discourse of rights is," as Tom Campbell (2006, 3) notes, "pervasive and popular in politics, law and morality. There is scarcely any position, opinion, claim, criticism or aspiration relating to social and political life that is not asserted and affirmed using the term 'rights.' Indeed, there is little chance that any cause will be taken seriously in the contemporary world that cannot be expressed with a demand for the recognition or enforcement of rights of one sort or another." But it is precisely this pervasiveness and widespread usage that is the problem and a potential obstacle to clear understanding. "Rights" is a rather loose signifier that is applied in ways that are not always consistent or even carefully delineated. Even experienced jurists, as Wesley Hohfeld (1920) noted nearly a hundred years prior to the current volume, have a proclivity to conflate different senses of the term and often times mobilize varied (if not contradictory) understandings of the word in the course of a decision or even a single sentence.

### 1.2.1 Definition

In order to redress this perceived deficiency and lend some terminological precision and clarity to the use of the term "rights," Hohfeld developed a classification schema, analyzing rights into four fundamental types, or what are commonly called the "Hohfeldian incidents." Although Hohfeld's work was initially developed in the context of law and therefore applies to the concept of legal rights, his typology has been successfully ported to, modified for, and used to explain moral and political rights as well.[2] The Hohfeldian incidents include the following four types of rights: privileges, claims, powers, and immunities. These four incidents are further organized into two groupings, the first two (privileges and claims) are considered "primary rights" (Hart 1961) or "first-order incidents" (Wenar 2005, 232), while the other two (powers and immunities) fall under the category of "secondary rights" (Hart 1961) or "second-order incidents" (Wenar 2005, 232). Explanations and analyses of the four incidents can be found throughout the literature on rights in works of social political philosophy (MacCormick 1982, Gaus and D'Agostino 2013), moral philosophy (Sumner 1987, Steiner 1994), and law (Hart 1961, Dworkin 1977, Raz 1986 and 1994). The following characterizations are taken from Leif Wenar's "The Nature of Rights" (2005):

**First Order Incidents** (1) Privileges (also called "liberty" or "license") describe situations where "A has a right to φ," or stated more precisely:

"A has a Y right to φ" implies "A has no Y duty not to φ."

(where "Y" is "legal," "moral," or "customary," and φ is an active verb) (Wenar 2005, 225).

As an illustration of this type of right, Wenar (2005) offers the following: "A sheriff in hot pursuit of a suspect has the legal right to break down the door that the suspect has locked behind him. The sheriff's having a legal right to break down the door implies that he has no legal duty not to break down the door" (225). The sheriff has a privilege (to break down the door) that overrides or "trumps" (which is the term utilized by Dworkin 1984) the obligation or duty that customarily applies in these situations, i.e., that one should not go around breaking down closed doors.

(2) Claims describe situations where "A has a right that B φ." "This second fundamental form of rights-assertion," as Wenar (2005, 229) explains, "often implies not a lack of a duty in the rightholder A, but the presence of a duty in a second party B."

"A has a Y right that B φ" implies "B has a Y duty to A to φ."

(where "Y" is "legal," "moral," or "customary," and φ is an active verb) (Wenar 2005, 229).

Again examples help clarify how this incident operates: "Your right that I not strike you correlates to my duty not to strike you. Your right that I help you correlates to my duty to help you. Your right that I do what I promised correlates to my duty to do what I promised" (229).

**Second Order Incidents**  The second order incidents can be understood and formulated as modification of the first order incidents. "We have," Wenar (2005, 230) explains, "not only privileges and claims, but rights to alter our privileges and claims, and rights that our privileges and claims not be altered." The former modification is called a "power"; the latter is called "immunity."

(3) Powers. The power modifies the first of the first order incidents—those propositions that take the form "A has a right to φ"—by imposing some restriction on oneself or another. "To have a power," Wenar (2005, 231) writes: "is to have the ability within a set of rules to alter the normative situation of oneself or another. Specifically, to have a power is to have the ability within a set of rules to create, waive, or annul some lower-order incident(s)." The power, therefore takes the following logical form:

"A has a power *if and only if* A has the ability to alter her own or another's Hohfeldian incidents" (Wenar 2015, 2).

As an example, Wenar offers the following rather odd description: "A ship's captain has the power-right to order a midshipman to scrub the deck. The captain's exercise of this power changes the sailor's normative situation: it imposes a new duty upon the sailor and so annuls one of his Hohfeldian privileges (not to scrub the deck)" (Wenar 2015, 2). Although a bit odd and not necessarily accessible to individuals lacking personal experience with naval operations, the example does provide an adequate illustration of what is typically involved in the Hohfeldian incident called "power."

(4) Immunities are modification of the second first-order incidents. "The immunity, like the claim," Wenar (2005, 232) explains, "is signaled by the form 'A has a right that B φ' (or, more commonly, '… that B not φ'). Rights that are immunities, like many rights that are claims, entitle their holders to *protection* against harm or paternalism." The immunity can, therefore, be described as such:

"B has an immunity *if and only if* A lacks the ability to alter B's Hohfeldian incidents" (Wenar 2015, 3).

As an illustration of the immunity, Wenar offers an example derived from US constitutional law: "The United States Congress lacks the ability within the Constitution to impose upon American citizens a duty to kneel daily before a cross. Since the Congress lacks a power, the citizens have an immunity. This immunity is a core element of an American citizen's right to religious freedom" (Wenar 2015, 3). Again the example is rather specific and limited to constitutional law and political freedom as formulated in the United States, but it does provide a sufficient illustration of how the concept of immunity operates.

There are two additional items to note in Hohfeld's typology. First, the four types of rights or incidents are all formulated in such a way that each one necessitates a correlative duty. If one has a right (a privilege, a claim, a power, or an immunity), it means that there is an associated duty imposed upon another who is obligated to respect that right. Or as Marx and Tiefensee (2015, 71) explain: "The 'currency' of rights would not be of much value if rights did not impose any constraints on the actions of others. Rather, for rights to be effective they must be linked with correlated duties." Hohfeld, therefore, initially presents the four incidents in terms of rights/duties pairs:

If A has a Privilege, then someone (B) has a No-claim.

If A has a Claim, then someone (B) has a Duty.

If A has a Power, then someone (B) has a Liability.

If A has an immunity, then someone (B) has a Disability.

This means that a statement concerning rights can be considered either from the side of the possessor of the right (the "right" that A has or is endowed with), which is a "patient oriented" way of looking at a moral, legal, or political situation; or from the side of the agent (what obligations are imposed on B in relationship to A), which considers the responsibilities of the producer of a moral, legal, or social/political action.

Second, although each of the incidents can by itself specify a particular right, the majority of rights are composed of a combination of more than one incident, or what Wenar (2005, 234) calls a "molecular right." A good example of this is property rights, specifically the right the owner of a piece of technology, like a computer, has concerning this object:

The "first-order" rights are your legal rights directly over your property—in this case, your computer. The privilege on this first level entitles you to use your computer. The claim correlates to a duty in every other person not to use your computer. The "second-order" rights are your legal rights concerning the alteration of these first-order rights. You have several powers with respect to your claim—you may waive the claim (granting others permission to touch the computer), annul the claim (abandoning the computer as your property), or transfer the claim (making the computer into someone else's property). Also on the second order, your immunity prevents others from altering your first-order claim over your computer. Your immunity, that is, prevents others from waiving, annulling, or transferring your claim over your computer. The four incidents together constitute a significant portion of your property right. (Wenar 2015, 4)

Because the Hohfeldian incidents either individually or in some combination describe most, if not all, situations and circumstances involving what are typically called rights, Wenar concludes that the Hohfeldian incidents are complete. "Any assertion of a right can be translated into an assertion about a single Hohfeldian incident, or into an assertion about a complex of incidents, or into a set of alternative assertions about such incidents. All rights are Hohfeldian incidents" (Wenar 2005, 235). The proof of this statement, Wenar admits, is inductive: "Our confidence in this inductive step will increase as we successfully explicate more and more rights with the Hohfeldian diagrams, and as we fail to find counterexamples. The reader may want to satisfy himself or herself that confidence in this inductive step is justified, and may wish to test the framework with more sample rights" (236). Though he is reasonably confident that the Hohfeldian incidents are complete and entirely sufficient for characterizing anything and everything that would be considered a right, this particular statement leaves the door open to the fact that it may be possible, at some point, to discover or identify a right that escapes the grasp of the Hohfeldian typology. Consequently, we will need to hold open the possibility that "robot rights" might, on the one hand, fit entirely within the typology, or, on the other hand, could provide the kind of counter example that would decrease Wenar's confidence.

### 1.2.2 Theories of Rights

Though Hohfeld's typology defines what rights are, it does not explain who has a right or why. This is the purview of theory, and there are two competing candidates concerning this issue. "One such theory," as Campbell

(2006, 43) explains, "is the 'will' (or 'choice' or 'power') theory. The other is the 'interest' (or 'benefit' or 'wellbeing') theory." Interest theories connect rights to matters of welfare. "Interest theorists maintain that individuals are capable of having rights if they possess certain interests, the value of which provides reason to impose duties on others" (Marx and Tiefensee 2015, 72). Although there is some variation in the way this has been articulated and formalized, "any doctrine classifiable as an Interest Theory of rights would," as Matthew Kramer (1998, 62) has explained, "subscribe to the following two theses":

(1) Necesssary but insufficient for the actual holding of a right by a person $X$ is that the right, when actual, preserves one or more of $X$'s interests.
(2) $X$'s being competent and authorized to demand or waive the enforcement of a right is neither sufficient nor necessary for $X$ to be endowed with that right.

"Will" theorists, by contrast, require that the subject of a right possess the authority and/or capacity to assert the privilege, claim, power, or immunity. "According to advocates of the Will or Choice Theory, only those individuals qualify as right-holders who have the ability to exercise a right. That is, only a person who can choose either to impose or waive the constraints that her right places on others' conduct can be seen to have rights" (Marx and Tiefensee 2015, 72). Once again there is some variability in the literature concerning how this particular theory is formulated. But all or most of them, as Kramer (1998, 62) argues, share these three principles:

(1) Sufficient and necessary for $X$'s holding of a right is that $X$ is competent and authorized to demand or waive the enforcement of the right.
(2) $X$'s holding of a right does not necessarily involve the protection of one or more of $X$'s interests.
(3) A right's potential to protect one or more of $X$'s interests is not sufficient per se for $X$'s actual possession of that right.

Although a great deal has been written about these two theories and their importance in the construction and development of rights discourse, we can, for the purposes of what follows, take note of two important consequences. First, comparing the common elements of these two theories, it is immediately evident that one of the major differences between Will Theory and Interest Theory is scope, that is, determinations regarding who can be included in the community of entities possessing rights and what is to be excluded as rights-less. Will Theory is designed to be more narrow, restrictive, or conservative on this account, while Interest Theory admits and can accommodate a wider range of entities. In other words, because Will Theory requires rights' holders to be "competent and authorized to demand or waive the enforcement of the right," it excludes from consideration human persons of diminished mental capacity, children, animals, the environment, etc. By contrast, "the Interest Theory can readily ascribe rights to children and to mentally incapacitated people (and indeed to animals, if a theorist wishes); it can acknowledge that the criminal law bestows individual rights on the people who are protected by the various statutes and other mandates that impose criminal duties; and it maintains that any genuine right does not have to be waivable and enforceable by the right-holder but can be waivable and enforceable by someone else instead" (Kramer 1998, 78).

Second, each theory has its advocates and champions. Matthew Kramer (1998), for instance, is a vehement advocate of the Interest Theory, arguing that it provides a more adequate formulation of the way that rights actually operate than the more restrictive Will Theory, which risks excluding entire populations of legitimate moral and legal subjects; while Hillel Steiner (1998, 234) defends Will Theory, arguing that it "offers a perfectly general account of what it is to have a right in any type of society" and that "the account it offers is distinctly more plausible than that afforded by the Interest Theory." The debate between these two factions is ongoing and appears to be unresolvable. "The long and unresolved historical contest between these two single-function theories," Wenar (2015, 238) concludes, "stretches back through Bentham (an interest theorist) and Kant (a will theorist) into the Dark Ages." And its most recent contest, which was staged in Kramer, Simmond, and Hillel's *A Debate Over Rights* (1998) ended in what Wenar (2015) and others have acknowledged as a "stalemate." For that reason, it would be impetuous to think that we can, in the course of this analysis, either resolve the debate once and for all or declare and defend the choice of one theory to the exclusion of the other. As with the polysemia that is involved with the word "robot," there is more to be gained by tolerating and explicitly accounting for differences in the theory of rights in order to be responsive to the diverse ways that rights have been mobilized in the existing literature.

This means not selecting between winners and losers in the debate but recognizing what each theory makes available, explicitly identifying when one is operationalized and asserted over the other, and ascertaining how this difference is able to supply critical perspective for better understanding what is gained and what is lost in the process of operationalizing one theoretical perspective as opposed to the other. We will, therefore, take the approach that Slavoj Žižek (2006a) calls "the parallax view." "Parallax" is the term Žižek deploys (in something of an appropriation and repurposing of the word's standard connotation) to name an irreducible difference in perspective that is not programmed for, nor determined to result in, some kind of final mediation or dialectical resolution. "The two perspectives," Žižek (2006a, 129) writes, "are irretrievably out of sync, so that there is no neutral language to translate one into the other, even less to posit one as the 'truth' of the other. All we can ultimately do in today's condition is to remain faithful to this split as such, to record it." With the two competing theories of rights, therefore, the truth does not reside on one side or the other or in their mediation through some kind of dialectical synthesis or hybrid construction (cf. Wenar 2015). The truth of the matter can be found in the shift of perspective from the one to the other. The critical task, therefore, is not to select the right theory of rights but to mobilize theory in the right way, recognizing how different theories of rights frame different problematics, make available different kinds of inquiries, and result in different possible outcomes.

## 1.3   Robot Rights or the Unthinkable

"The notion of robot rights," as Seo-Young Chu (2010, 215) points out, "is as old as is the word 'robot' itself. Etymologically the word 'robot' comes from the Czech word 'robota,' which means 'forced labor.' In Karel Čapek's 1921 play *R.U.R.*, which is widely credited with introducing the term 'robot,' a 'Humanity League' decries the exploitation of robot slaves—'they are to be dealt with like human beings,' one reformer declares—and the robots themselves eventually stage a massive revolt against their human makers." For many researchers and developers, however, the very idea of "robot rights" is simply unthinkable. "At present," Sohail Inayatullah and Phil McNally (1988, 123) once wrote, "the notion of robots with rights is unthinkable whether one argues from an 'everything is alive' Eastern perspective or 'only man is alive' Western perspective." But why? How and why is the mere idea of "robot rights" inconceivable? What does it mean, when the mere thought of something is already decided and declared to be unthinkable?

### 1.3.1 Ridiculous Distractions

Considering the impact of his earlier (1988) essay with Phil McNally, Inayatullah (2001, 1) provided the following reflection concerning the question of robot rights:

> Many years ago in the folly of youth, I wrote an article with a colleague titled, "The Rights of Robots." It has been the piece of mine most ridiculed. Pakistani colleagues have mocked me saying that Inayatullah is worried about robot rights while we have neither human rights, economic rights or rights to our own language and local culture—we have only the "right to be brutalized by our leaders" and Western powers. Others have refused to enter in collegial discussions on the future with me as they have been concerned that I will once again bring up the trivial. As noted thinker Hazel Henderson said, I am happy to join this group—an internet listserv—as long as the rights of robots are not discussed. But why the ridicule and anger?

As Inayatullah reports from his own experience, the mere idea of robot rights—not even advocacy for rights but the suggestion that the question might require some careful consideration—has produced anger, mockery, and ridicule in colleagues and noted thinkers. In the face of this seemingly irrational response, Inayatullah asks a simple but important question, "why the ridicule and anger?" In response to this, he calls upon the work of Christopher Stone, who once pointed out that any extension of rights to previously excluded populations—like animals or the environment—is always, at least initially, considered unthinkable and the target of ridicule. "Throughout legal history," Stone (1974, 6–7) writes, "each successive extension of rights to some new entity has been, theretofore, a bit unthinkable. ... The fact is, that each time there is a movement to confer rights onto some new 'entity,' the proposal is bound to sound odd or frightening or laughable." This mode of response is, for example, evident in the editors' note appended to the beginning of a reprint of Marvin Minsky's "Alienable Rights": "Recently we heard some rumblings in normally sober academic circles about robot rights. We managed to keep a straight face as we asked Marvin Minsky, MIT's grand old man of artificial intelligence, to address the heady question" (Minsky 2006, 137).[3] From the outset, the question of robot rights is presumed to be ludicrous, so much so that one struggles to keep a straight face in the face of the mere idea. Whatever the exact reason for this kind of reaction—whether it be fear, resentment, prejudice, etc.[4]—the effect of ridicule is to discredit or to dismiss the very idea.

Similar responses have been registered and documented elsewhere in the literature. In 2006, the UK Office of Science and Innovation's Horizon Scanning Centre commissioned and published a report consisting of, as Raya Jones (2016, 36) explains, "246 summary papers that predict emerging trends in science, health, and technology." The papers, or "scans" as Sir David King referred to them, were "aimed at stimulating debate and critical discussion to enhance government's short and long term policy and strategy" (BBC News 2006). Despite the fact that these scans (which had been contracted to and developed by Ipsos MORI in 2006) addressed a number of potentially controversial subjects, only one was "singled out to make headlines" (Jones 2016, 36). Ipsos MORI titled the article: "Robo-rights: Utopian Dream or Rise of the Machines?" (which was called "Robot Rights: Utopian Dream or Rise of the Machines?" in Winfield 2007 and "Robo-rights: Utopian Dream or Rise of the Machine?" in Bensoussan and Bensoussan 2015). What caught everyone's attention was the following statement: "If artificial intelligence is achieved and widely deployed (or if they can reproduce and improve themselves) calls may be made for human rights to be extended to robots" (Marks 2007; Ipsos MORI 2006). The statement was short and the report itself, which was "a little over 700 words" (Winfield 2007), did not supply much in terms of actual details or evidence. But it definitely had impact. It was covered by and widely reported in the popular press, with attention grabbing headlines like "UK report says robots will have rights" (Davoudi 2006) and "Robots could demand legal rights" (BBC News 2006).

The response from researchers and experts in the field was highly critical.[5] As Robert M. Geraci (2010, 190) explained: "The Ipsos MORI document was reviled by some scientists in the UK. Rightly criticizing the study for its minimal research documentation, Owen Holand [SIC], Alan Winfield, and Noel Sharkey claim the Ipsos MORI document directs attention away from the real issue, especially military robots and responsibility for autonomous robot-incurred death or damage (Henderson 2007). The scientists gathered at the Dana Centre in London in April of 2007 to share their ideas with the public." This account of the sequence of events is not entirely accurate.

Holland, Sharkey, and Winfield—along with Kathleen Richardson, whose name is not, for some reason, included in the published news stories—shared their comments about the Ipsos MORI report one day before the Dana Centre event during a press briefing at the Science Media Centre (SMC). In their remarks to the press, Holland, Sharkey, and Winfield—who together with Frank Burnett were co-PIs on the "Walking With Robots" research program, a multi-year public information project funded by the UK's Engineering and Physical Sciences Research Council (2005)—not only criticized the report but reportedly dismissed the concept of robot rights as a "waste of time and resources" (Geraci 2010, 190) and a "red herring that has diverted attention from more pressing ethical issues" (Henderson 2007). As Holland explained, the report was "very shallow, superficial, and poorly informed. I know of no one within the serious robotics community who would use that phrase, 'robot rights.'" (Henderson 2007). And according to Sharkey "the idea of machine consciousness and rights is a distraction, it's fairy tale stuff. We need proper informed debate, about the public safety about for instance the millions of domestic robots that are predicted to be arriving in the next few years" (Henderson 2007 and Marks 2007).

By all accounts, the report was insufficient, perhaps necessarily so, since it was not a peer-reviewed scientific study but a short thought-piece designed to motivate public discussion. There is, however, a bit of oversteer in these critical responses. In addition to directing criticism at the document and its perceived deficiencies, the respondents also characterized the entire subject of robot rights as a fantastic distraction from the serious work that needs to be done and therefore something no serious researcher would think about, let alone say out loud. In all fairness, it is entirely possible that the remarks were motivated by a well-intended effort to try to direct public attention to the "real problems" and away from dramatic media-hype and futurist speculation. And this is undoubtedly one of the important and necessary objectives of science education. But even if we grant this, the comments demonstrate something of a dismissive response to the question of robot rights and its relative importance (or lack thereof) in current research programs. Even when taking a more measured and less polemic approach, as Winfield did in a blog post from February of 2007, the "rights question" is, if not dismissed, then indefinitely delayed by being put aside or pushed off into an indeterminate future.

Ok, let's get real. Do I think robots will have (human) rights within 20–50 years? No, I do not. Or to put it another way, I think the likelihood is so small as to be negligible. Why? Because the technical challenges of moving from insect-level robot intelligence, which is more or less where we are now, to human-level intelligence are so great. Do I think robots will ever have rights? Well, perhaps. In principle I don't see why not. Imagine sentient robots, able to fully engage in discourse with humans, on art, philosophy, mathematics; robots able to empathi[z]e or express opinions; robots with hopes, or dreams. Think of Data from Star Trek. It is possible to imagine robots smart, eloquent and persuasive enough to be able to argue their case but, even so, there is absolutely no reason to suppose that robot emancipation would be rapid, or straightforward. After all, even though the rights of man as now generally understood were established over 200 years ago, human rights are still by no means universally respected or upheld. Why should it be any easier for robots? (Winfield 2007).

In this post, Winfield does not dismiss the concept of robot rights per se, but comes at it directly. In the short term (20–50 years), it appears to be, if not impossible, then at least highly improbable. In the long run, it may be something that is, in principle at least, not impossible. This possibility, however, is projected into a distant hypothetical future on the basis of the more exclusive Will Theory of rights, i.e., that robots would be "smart, eloquent and persuasive enough to be able to argue their case." The question concerning robot rights, therefore, is not "unthinkable"; it is just unlikely and practically not worth the effort at this particular point in time. As Holland had succinctly stated it during the press briefing, "it's really premature I think to discuss robot rights" (Randerson 2007).

A similar decision is operationalized in the official response issued by the Foundation for Responsible Robotics to the European Parliament's recently published recommendations to the European Commission concerning robots and AI (Committee on Legal Affairs 2016): "Some of the wording in the document is disappointingly based on Science Fiction. Notions of robot rights and robot citizenship are really a distraction from the main thrust of the document that only served to catch the wrong type of media attention" (Sharkey and Fosch-Villaronga 2017). Likewise, Luciano Floridi (2017, 4) suggests that thinking and talking about the "counterintuitive attribution of rights" to robots is a distraction from the serious work of philosophy. "It may be fun," Floridi (2017, 4) writes, "to speculate about such questions, but it is also distracting and irresponsible, given the pressing issues we have at hand. The point is not to decide whether robots will qualify someday as a kind of persons, but to reali[z]e that we are stuck within the wrong conceptual framework. The digital is forcing us to rethink new solutions for new forms of agency. While doing so we must keep in mind that the debate is not about robots but about us, who will have to live with them, and about the kind of infosphere and societies we want to create. We need less science fiction and more philosophy." According to Floridi, it might be fun to speculate about things like robot rights, but this kind of playful thinking is a distraction from the real work that needs to be done and quite possibly is an irresponsible waste of time and effort that should be spent on more serious philosophical endeavors.

### 1.3.2 Justifiable Exclusions

In other situations the question of robot rights is not immediately ridiculed or dismissed; it is given some brief attention only to be bracketed or carefully excluded as an area that shall not be given further thought. One of the best illustrations of this is deployed by Michael Anderson, Susan Leigh Anderson, and Chris Armen (2004) in the agenda-setting paper that launched the new field of machine ethics. For Anderson et al., machine ethics (ME) is principally concerned with the consequences of the decisions and behaviors of machines toward human beings. For this reason, Anderson et al. immediately differentiate ME from two related efforts: "Past research concerning the relationship between technology and ethics has largely focused on responsible and irresponsible use of technology by human beings, with a few people being interested in how human beings ought to treat machines." ME is introduced and defined by being differentiated from two other ways of considering the relationship between technology and ethics. The first is computer ethics, which is concerned, as Anderson and company correctly point out, with questions of human action through the instrumentality of computers and related information systems. In clear distinction from these efforts, machine ethics seeks to enlarge the scope of moral agents by considering the ethical status and actions of machines. As Anderson and Anderson (2007, 15) describe it in a subsequent publication, "the ultimate goal of machine ethics, we believe, is to create a machine that itself follows an ideal ethical principle or set of principles." The other exclusion involves situations regarding rights, or "how human beings ought to treat machines." This also does not fall under the purview of ME, and Anderson, Anderson, and Armen explicitly mark it as something to be set aside by their own endeavors. Although the "question of whether intelligent machines should have moral standing," Susan Leigh Anderson (2008, 480) writes in another article, appears to "loom on the horizon," ME deliberately and explicitly pushes this issue to the margins, leaving it for others to consider. Here we can perceive how that which is marginalized appears; it makes an appearance in the text only insofar as it is set aside and not given further consideration. It appears, as Derrida (1982, 65) describes it, by way of the trace of its erasure. This setting-aside or exclusion leaves a residue in the text—a mark within the text of something being excluded—that remains legible and can be read.

A similar decision is deployed in Wendell Wallach and Colin Allen's *Moral Machines* (2009), and is clearly evident in their choice of the term "artificial moral agent," or AMA, as the protagonist of the analysis. This term immediately focuses attention on the question of responsibility. Despite this exclusive concern, however, Wallach and Allen (2009, 204–207) do eventually give brief consideration to the issue of rights. Extending the concept of legal responsibility to AMAs is, in Wallach and Allen's opinion, something of a no-brainer: "The question whether there are barriers to designating intelligent systems legally accountable for their actions has captured the attention of a small but growing community of scholars. They generally concur that the law, as it exists, can accommodate the advent of intelligent (ro)bots. A vast body of law already exists for attributing legal personhood to nonhuman entities (corporations). No radical changes in the law would be required to extend the status of legal person to machines with higher-order faculties, presuming that the (ro)bots were recognized as responsible agents" (Wallach and Allen 2009, 204). According to Wallach and Allen's estimations, a decision concerning the legal status of AMA's should not pose any significant problems. Most scholars, they argue, already recognize that this is adequately anticipated by and already has a suitable precedent in available legal and judicial practices, especially as it relates to the corporation.

What is a problem, in their eyes, is the flip side of legal responsibility—the question of rights. "From a legal standpoint," Wallach and Allen continue, "the more difficult question concerns the rights that might be conferred on an intelligent system. When or if future artificial moral agents should acquire legal status of any kind, the question of their legal rights will also arise" (Wallach and Allen 2009, 204). Although noting the possibility and importance of the question, at least as it would be characterized in legal terms, they do not pursue its consequences very far. In fact, they mention it only to defer it to another kind of question—a kind of investigative bait and switch: "Whether or not the legal ins and outs of personhood can be sorted out, more immediate and practical for engineers and regulators is the need to evaluate AMA performance" (Wallach and Allen 2009, 206). Consequently, Wallach and Allen conclude *Moral Machines* by briefly gesturing in the direction of a consideration of robot rights only to refer this question back to the issue of agency and performance measurement. In this way, then, they briefly address the rights question only to immediately recoil from the complications it entails, namely, the persistent philosophical problem of having to sort out "the legal ins and outs of personhood." Although not simply passing over the question of robot rights in silence, Wallach and Allen, like Anderson et al., only mention it in order to postpone or otherwise exclude the issue from further consideration.

In other cases, the exclusion is not even explicitly marked or justified; it is simply declared and enacted. The EURON *RoboEthics Roadmap* (Veruggio 2006), for instance, does not have anything to say (not even a negative exclusion) about the social standing or rights of robots. As Yueh-Hsuan Weng, Chien-Hsun Chen, and Chuen-Tsai Sun (2009, 270) report, "the *Roboethics Roadmap* does not consider potential problems associated with robot consciousness, free will, and emotions" but takes an entirely human-centric approach. The *Roadmap*, then, appears to pursue what is a rather reasonable path. The scope of the report is limited to a decade, and, for this reason, the authors of the document set aside the more speculative questions concerning future achievements with robot capabilities. Though robot rights is thinkable in principal, in practice it is left off the table as a matter of setting some practical limits to the investigation.

> In terms of scope, we have taken into consideration—from the point of view of the ethical issues connected to Robotics—a temporal range of a decade, in whose frame we could reasonably locate and infer—on the basis of the current State-of-the-Art in Robotics—certain foreseeable developments in the field. For this reason, we consider premature—and have only hinted at—problems inherent in the possible emergence of human functions in the robot: like consciousness, free will, self-consciousness, sense of dignity, emotions, and so on. Consequently, this is why we have not examined problems—debated in the literature—like the need to consider robot as our slaves, or the need to guarantee them the same respect, rights and dignity we owe to human workers. (Veruggio 2006, 7)

In this document, the question of rights is deliberately set aside as a matter that is premature insofar as it requires advancements in robotics (i.e., consciousness, free will, emotions, etc.) that remain futuristic, debatable, and outside the scope of the study's ten-year window of opportunity.

Not surprisingly, a similar sounding decision is instituted by Gianmarco Veruggio (the project coordinator of the EURON *Roboethics Roadmap*) and Fiorella Operto, in a subsequent contribution to the *Springer Handbook of Robotics*:

> In terms of scope, we have taken into consideration—from the point of view of the ethical issues connected to robotics—a temporal range of two decades, in whose frame we could reasonably locate and infer—on the basis of the current state-of-the-art in robotics—certain foreseeable developments in the field. For this reason, we consider premature—and have only hinted at—problems related to the possible emergence of human qualities in robots: consciousness, free will, self-consciousness, sense of dignity, emotions, and so on. Consequently, this is why we have not examined problems—debated in some other papers and essays—like the proposal to not behave with robots like with slaves, or the need to guarantee them the same respect, rights, and dignity we owe to human workers. Likewise, and for the same reasons, the target of roboethics is not the robot and its artificial ethics, but the human ethics of the robots' designers, manufacturers, and users. Although informed about the issues presented in some papers on the need and possibility to attribute moral values to robots' decisions, and about the chance that in the future robots might be moral entities like—if not more so than—human beings, the authors have chosen to examine the ethical issues of the human beings involved in the design, manufacturing, and use of the robots. (Veruggio and Operto 2008, 1501)

Here again, Veruggio (who is credited with initially introducing and defining the project of *roboethics*) and Operto limit consideration to the near-term—two decades. They therefore explicitly set aside and postpone any consideration of robots as either independent agents or patients and focus their attention on a kind of elaboration of computer ethics, that is, the ethical issues encountered by human beings in the design, manufacture, and use of robots.

A similar future-oriented disclaimer is deployed in Michael Nagenborg et al.'s critical investigation of robot regulation across Europe. In this text, the authors consider a number of issues regarding "responsibility and autonomous robots," but they set aside the question of rights as not pertinent to the matter at hand: "In the context of this paper we therefore assume that artificial entities are not persons and not the bearers of individual, much less civil, rights. This does not imply that in a legal and ethical respect we could not grant a special status to robots nor that the development of artificial persons can be ruled out in principle. However, for the present moment and the nearer future we do not see the necessity to demand a fundamental change of our conception of legality. Thus, we choose a human-centred approach" (Nagenborg et al. 2008, 350). Nagenborg and company can safely and justifiably make this assumption, which to their credit they recognize and identify as an "assumption," because robots are widely regarded as things that are not "bearers of individual, much less civil, rights" (2008, 350). This could, the authors recognize, change at some point in the future, and that change, in distinction to what Winfield (2007) had argued, could be made and justified in term of an Interest Theory insofar as we would decide to "grant ... special status to robots." But that future is so distant that it is not even worth serious consideration at this particular moment.

In other situations, exclusion occurs not as a deliberate postponement but as a product of the means of inclusion. On the legal side, for instance, the matter of "robot rights" is already effectively disabled by the way we have decided to incorporate robots into the law and how that mode of incorporation differs from another, similarly situated entity. As Gunther Teubner (2006, 521) explains:

Law is opening itself for the entry of new juridical actors—animals and electronic agents. The difference in outcomes, however, are striking. ... Animal rights and similar constructs create basically defensive institutions. Paradoxically, they incorporate animals in human society in order to create defenses against destructive tendencies of human society against animals. The old formula of social domination of nature is replaced by the new social contract with nature. For electronic agents, the exact opposite is true. Their legal personification, especially in economic and technological contexts, creates aggressive new action centers as basic productive institutions. Here, their inclusion into society does not protect the new actors—just the opposite. It is society that needs to defend itself against the new actors.

Both robots and animals can be considered "the excluded other" of human social institutions, especially legal institutions. The way these two previously marginalized entities come to be included in law, however, is remarkably different. Animals—at least some species of animals—now have rights that are legally recognized in an effort to defend the welfare and interests of these vulnerable entities from "the destructive tendencies of human society." Robots, or what Teubner calls "electronic agents" (which is already a mark of difference, insofar as animals are not describe as "agents" but occupy the place of "patient"), come to be included in legal matters for the opposite reason. Robots figure in law not for the sake of protecting the rights of the machine from human exploitation and misuse but for the sake of defending human society from potentially disastrous robot decision-making and action. Animals have rights; robots have responsibilities. It is, therefore, not the case that robots are simply excluded from the law, but that the means of their inclusion is accomplished in such a way that already renders the question of rights inoperative and effectively unthinkable.

### 1.3.3 Literal Marginalization

In still other situations, the question of robot rights is not excluded per se but included by being pushed to the margins of proper consideration. What is marginal is noted and given some attention, but it is deliberately marked as something located on the periphery of what is determined to be really important and worthy of investigation. Indication of the marginal position or at least the peripheral status of the "rights question" can be found in Wendell Wallach and Peter Asaro's collection of foundational texts in robot and machine ethics (2017). Out of the thirty-five essays that are assembled in this volume, which is intended "to provide a roadmap of core concerns" by bring together a "careful selection of articles that have or will play an important role in ongoing debates" (Wallach and Asaro 2017, xii), only one essay, the final one, takes up and engages the question of rights. The position of this text within the sequence of the chapters, along with the introductory explanation provided by the editors, situate the inquiry into robot rights as a kind of future-oriented epilogue:

Perhaps the most futuristic and contested question in robot ethics is whether robots may someday have the capacity or capability to be the bearer of rights, and in virtue of what criteria will they deserve those rights? Gunkel (2014) defends the view that there may be robots deserving of the legal and moral status of having rights, and challenges the current standards of evaluating which entities (corporations, animals, etc.) deserve moral standing. Such futuristic legal questions underscore the way in which artificial agents presently serve as a thought experiment for legal theorists and a vehicle for reflections on the nature of legal agency and rights. (Wallach and Asaro 2017, 12)

In pointing this out, I am not complaining. It was both an honor and privilege to have this text included in the project. But the manner of its inclusion—the fact that it is the sole essay addressing the question of robot rights and that it comes "at the end of the story"—indicates the extent to which the "rights question" has been and remains a kind of marginalized afterthought in the literature of robot and machine ethics.

This literal maginalization is also evident in Patrick Lin et al.'s *Robot Ethics*, which has now appeared in two versions. In the first iteration, *Robot Ethics* (Lin, Abney, and Bekey 2012), the question concerning robot rights is explicitly addressed in the book's final section, "VII Rights and Ethics." This section consists of three essays from Rob Sparrow (2012), Kevin Warwick (2012), and Anthony F. Beavers (2012), and it is introduced with the following editorial note:

> The preceding chapter 18 examined the ethics of robot servitude: Is it morally permissible to enforce servitude on robots, sometimes termed "robot slavery"? But to call such servitude "slavery" is inapt, if not seriously misleading, if robots have no will of their own—if they lack the sort of freedom we associate with moral personhood and moral rights. However, could robots someday gain what it takes to become a rights holder? What exactly is it that makes humans (but not other creatures) here on Earth eligible for rights? Is there a foreseeable future in which robots will demand their own "Emancipation Proclamation"? (Lin, Abney, and Bekey 2012, 299)

Like Wallach and Asaro, Lin et al. conclude *Robot Rights* (version 1, 2012) by giving consideration to a set of questions involving somewhat distant matters that could arise in the foreseeable future. "Robot rights," although not a central concern of *Robot Ethics* at the present time, is added on at the end as a kind of indication of "future directions" for possible research: "Thus, after studying issues related to programming ethics and specific areas of robotic applications, in part VII our focus zooms back out to broader, more distant concerns that may arise with future robots" (Lin, Abney, and Bekey 2012, 300). But in the second version of the book, *Robot Ethics 2.0* (Lin, Jenkins, and Abney 2017), this material is virtually absent. In this "major update" and "second generation" of *Robot Ethics*, what had been marginal in the first version—"marginal" insofar as it was a distant possibility situated at the end of the text—is now effectively pushed out of the book altogether.[6] This remarkable absence—the lack of something that had been present—is only able to be identified through its *trace* (Derrida 1982, 65), that is, the withdrawal of something that had previously been there, such that its lack of presence now leaves a mark or trace of its removal.[7]

### 1.3.4 Exceptions that Prove the Rule

Finally, there are some notable exceptions. But these exceptions, more often than not, prove the rule. In 1985, Robert A. Freitas Jr. published a rather prescient note in the journal *Student Lawyer*. Although considered to be one of the earliest documented attempts to raise the "questions of 'machine rights' and 'robot liberation'" (Freitas 1985, 54), this short article has had less than twenty citations in the past thirty-five years. This may be the result of: the perceived status (or lack thereof) of the journal, which is not a major venue for peer-reviewed research, but a magazine published by the student division of the American Bar Association; a product of some confusion concerning the essay's three different titles[8]; the fact that the article is not actually an article but an "end note" or epilogue; or simply an issue of timing, insofar as the questions Freitas raises came well in advance of robot ethics or even wide acceptance of computer ethics. In any event, this initial effort to situate the question concerning robot rights has been and remains on the periphery of scholarship, occupying a position that is, quite literally, an endnote.

Another notable exception is Blay Whitby's "Sometimes it's Hard to be a Robot: A Call for Action on the Ethics of Abusing Artificial Agents" from 2008. Although Whitby does not take up the question of robot rights (the term "rights" does not appear in his essay), his concern is with the potential for robot abuse.[9] "This is a call," Whitby (2008, 326) explains, "for informed debate on the ethical issues raised by the forthcoming widespread use of robots, particularly in domestic settings. Research shows that humans can sometimes become very abusive towards computers and robots particularly when they are seen as human-like and this raises important ethical issues." Whitby's objective is modest. He only seeks to open up the issue and initiate debate about the possible ethical consequences of future robot abuse. As such, and by deliberate design, he does not portend to offer any practical solutions or answers, which is something for which he is criticized by Thimbleby (2008). He only seeks to get the conversation started in an effort to begin the task of thinking about these potential problems. But even raising the question—even asking that roboticists, ethicists, and legal scholars entertain the question of robot abuse and begin to devise ethical standards in the interest of dealing with these potential complications—was perceived as being premature and unnecessary. As Thimbleby (2008, 341) writes (rather acerbically):

Whitby has not convinced this reader that robotism is distinct from any other technology-oriented ethics. He has not provided persuasive or provocative case studies, e.g., "it's not the gun that kills but the user" that arise frequently in other technoethical discussions. This reader does not share his sentimental interpretation of "abuse"; robots are not kittens. Robots are easy to build and mass produce; there are no grounds, at least as presented, to justify Whitby's sentimental title—"sometimes it's hard to be a robot ..." (Aw, I feel sorry for them already.) I imagine, if robots had any notion of what "hard" meant that they'd say it was easy enough to be a robot ... In conclusion, then, it should be clear that we disagree with Whitby: there is no urgency to define robotism or any ethics of robots. But as nothing is certain, we should end on a careful note: there is no urgency until the robots themselves start to ask for it.

Thimbleby, it should be noted, does not simply exclude consideration of robot rights *tout court* but is careful to recognize that if this arguably absurd idea were at some point in the future to become possible, it could only transpire in terms of the Will Theory of rights. The robots themselves will need to ask for it. Consequently, the difference between Whitby's and Thimbleby's positions is one that is ultimately grounded in the two different theories of rights. Whitby mobilizes an Interest Theory insofar as he advocates a consideration of the problems and consequences of robot abuse irrespective of whether the robot is capable (or not) of demanding that it not be abused. Thimbleby deploys a Will Theory approach, arguing that until the robots are able to ask for consideration or respect, we simply do not need to be concerned about it. A similar argument is made by Keith Abney (2012, 40), who asserts an exclusive privileging of the Will Theory as the basis for drawing the following conclusion: "we can safely say that, for the foreseeable future, robots will have no rights." Finally, even though Whitby's article has been referenced just under forty times in the literature, its main proposal, namely that we take up the question of robot abuse and consider some level of legal/moral protections for artifacts, is only directly addressed in a small fraction of these subsequently published texts (Riek et al. 2009, Coeckelbergh 2010 and 2011, and Dix 2017).

Another seemingly notable exception is Alain and Jérémy Bensoussan's *Droit des Robots* (2015), a book written by two attorneys involved in helping to define the emerging field of robot law. The French word *droit/droits* is, however, a bit tricky and can be translated as either "law" or "rights." *Droit*, in the singular and written either with or without the definite article (*le droit*), is typically translated as "law," as in *le droit penal* or "the penal law." The plural *droits*, especially when it is written with the definite article *les droits*, is typically translated "rights," as in the UN's *La Déclaration universelle des droits de l'homme* [The Universal Declaration of Human Rights]. But there is some equivocation concerning this matter, which makes the choice between "law" and "rights" something that is often dependent on context.

The title of Bensoussan and Bensoussan's book, for instance, would be translated as "Robot Law" and not "Right of Robots." And this translation is consistent with and verified by the contents of the text, which endeavors to articulate a new framework for responding to the legal challenges and opportunities of robots. In the course of their argument, however, Bensoussan and Bensoussan introduce a charter, composed of ten articles, that they call *La Chartes Des Droits Des Robots*. In this case, the word *droits* could be translated as "laws," "rights," or both, because the contents of the charter concern not only legal responsibilities for the manufacture, deployment, and use of robots, but also stipulate the legal status of robotic artifacts: "Article 2—Robot Person: A robot is an artificial entity endowed with its own legal personality: the robot personality. The robot has a name, an identification number, a state and a legal representative, which may be a natural person or a legal entity" (Bensoussan and Bensoussan 2015, paragraph 219; my translation). With this second article, Bensoussan and Bensoussan make the case for recognizing robots as a new kind of legal person, like a corporation. This is necessary, they argue, to contend with the unique social and legal challenges that robots present to current judicial systems. In other words, and unlike the vast majority of Anglophone researchers, scholars, and critics, Bensoussan and Bensoussan's work recognizes the difficulty of formulating laws [*droits*] that apply to robots without giving some consideration to the legal standing and rights [*les droits*] of the entity to which these laws would apply. And the one legal right that they believe is paramount for robots is the right to privacy.[10] This is stipulated in the third article: "Personal data stored by a robot are subject to the Data Protection and Freedom regulations. A robot has the right to the respect of its dignity limited to the personal data conserved" (Bensoussan and Bensoussan 2015, paragraph 219; my translation).

It should be noted, however, that the motivation and objective for extending legal protections to the robot remains focused on the human user. Here is how Alain Bensoussan explained it in an interview from 2014:

There is the question of duties, but also of rights. Such as the right to privacy. Indeed, the robot, in its interaction with the elderly or robots who "work" with autistic children, acquires information about their health and privacy. For example, the robot is able to tell a person with Alzheimer's disease "This is going to be your grandson's birthday" or to an autistic child "Your brother has arrived," or to an elderly person "Your granddaughter is here." And this intimacy must be protected by protecting the memory of the robot. (Bensoussan 2014, 13; my translation)

Consequently, Bensoussan and Bensoussan (2015) recognize the importance of explicitly stipulating the social and legal status of the robot and entertain questions regarding both robot law [*droit des robots*] and robot rights [*les droits des robots*], or at least one particular claim, the right to privacy. But for all its promise to think and deal with the question of robot rights, Bensoussan and Bensoussan's *Droit des Robots* is still restricted to an anthropocentric focus and frame of reference. The one right that the robot would have and would need to have—the right to privacy—is provided not for the sake of the robot but to protect the human beings involved with the robot. It is, therefore, a kind of indirect and instrumentalist way of thinking of the rights of others. Consequently, a robot's rights are not unthinkable, but their consideration is tightly constrained by, and limited to, the rights that belong to its human users.

## 1.4 Summary

Investigating the opportunities and challenges of robot rights is already complicated by both linguistic and conceptual difficulties. Linguistically, the term "robot" admits of a wide range of different and not necessarily compatible definitions, characterizations, and understandings. Like the word "animal," which Derrida (2008, 32) finds both compelling and astonishing—"Animal, what a word!"—the word "robot" is also a rather indeterminate and slippery signifier for a set of ideas and concepts that differ across texts, research projects, and investigative efforts. It is, therefore, and already from the beginning, "a question of words" and "an exploration of language" (Derrida 2008, 33). Even the most general and seemingly uncontroversial of characterizations—the sense-think-act paradigm—remains insufficient to capture and/or represent the full range and complexity that is already available in the current literature. And this "literature" includes not just academic, technical, and commercial publications but decades of science fiction as presented in print (i.e., short stories, novels, novellas, and comic books), stage plays and performances, and other forms of media (i.e., radio, cinema, and television); which in this particular case has determined many aspects of this thing called "robot"—or perhaps more accurately stated, these diverse and differentiated artifacts that have come to be designated and identified with the word "robot"—well in advance of actual efforts in science and engineering. In response to this, it has been decided not to make the usual definitive decision but to attend to and learn to take responsibility for this polysemia. That is, rather than proceeding by way of the "tried and true" approach to defining terminology and issuing a declaration that takes the form of something like, "In this study, we understand 'robot' to be …," we will endeavor to account for, to tolerate, and even to facilitate this semantic diversity and plurivocity. The word "robot," then, will function as a kind of semi-autonomous, linguistic artifact that is always and already on the verge of escaping from our oversight and control, such that the task before us is not to limit opportunity in advance of the investigation but to account for the full range of significations that the term "robot" puts into play and makes available.

A different but not necessarily unrelated terminological problem occurs with the word "rights." Unlike "robot," the term "rights" does have a rather strict and formal characterization. The Hohfeldian incidents parse the term into four distinct and interrelated elements: privileges, claims, powers, and immunities. For most moral philosophers and jurists, as Wenar (2005, 2015) suggests, these four incidents—either alone or in some combination—provide what is typically considered to be a complete formulation of and account for what we understand by the term "rights." Where things get complicated is in deciding who or what can or should have a share in these four incidents. Who or what, in other words, can have rights? As we have seen, there are two competing theories for addressing this question: the Interest Theory and the Will Theory. The latter sets a rather high bar, requiring that the subject of a right possess the authority and/or capacity to assert, on its own, a privilege, claim, power, and/or immunity that would need to be respected by others. The former provides for a much lower threshold, making it possible for one to decide rights on behalf of the interests of others.

The choice of theory makes a significant difference. Under the Will Theory, for instance, it would be virtually impossible for an animal or the environment to have any stake in the four incidents, and this insofar as they lack λόγος—that influential ancient Greek word that means "word" and can be translated as "speech," "language," "logic," and "reason"—and the ability to articulate and assert their rights. The Interest Theory, however, proceeds otherwise. It recognizes, following Bentham (2005, 283), that the operative question is not "Can they speak?" but "Can they suffer?" This other way of proceeding[11] would institute an entirely different way of deciding between who or what has rights and who or what does not. But this decision between the two theories is itself already a moral determination, for it already makes a decisive cut in the fabric of being separating *who* can be a moral subject from *what* remains a mere object situated outside the proper confines of moral concern. And this proto-moral determination—this "moral decision before moral decision" or this "ethics of ethics"—will be and will remain a subject of the investigation. Ultimately, because a decision concerning these two competing theories of rights remains unresolved and debated, we will need to keep both in play by using a strategy that Žižek calls "the parallax view," which is a way of perceiving and taking account of the difference that is available in the shift from the one theoretical perspective to the other.[12] What is important, therefore, is not selecting the "right theory" of rights at the beginning and sticking with it. What is important is developing the capability to call out and identify which theory is in play in which particular consideration of rights and how each of these frame distinctly different ways of responding to the question "Can and should robots have rights?"

The terms "robot" and "rights" are already complicated enough, but once you put the two words together, you get something of an allergic reaction. "Robot rights" is for many theorist and practitioners simply unthinkable, meaning that it is either unable to be thought, insofar as the very concept strains against common sense or good scientific reasoning; or is to be purposefully avoided as something that must not be thought—i.e., as a kind of prohibited idea or blasphemy that would open a Pandora's box of problems and therefore must be suppressed or repressed (to use common psychoanalytical terminology). Whatever the reason, there is something of a deliberate decision and concerted effort not to think—or at least not to take as a serious matter for thinking—the question of robot rights. The very idea is openly mocked as ridiculous and laughable; it is cited for engaging in pointless speculation that has the effect of distracting one from the true and serious work that needs to be done; or the subject is pushed aside and marginalized as a kind of appendix or sidebar that could be of interest in the future, but which (for now at least) does not really need serious attention. This is, however, precisely the reason why we must think the unthinkable—to challenge these declarations, assumptions, and orthodoxies; and this needs to transpire for at least two reasons.

First, any dogmatic declaration of this kind should already make us uneasy and suspicious. The fact that this thought—the mere thought that is made available in the association of the two words "robot" and "rights"—has been declared "unthinkable" should give one pause and trigger critical questions like: Who says? Who gets to decide in advance what we can or cannot think? And, perhaps more revealing, what values and assumptions are being protected through this kind of proscription and prohibition? As Barbara Johnson (1981, xv) reminds us, the task of critical thinking is to read backward from what seems obvious, self-evident, and natural, to show how these things already have a history that determines what becomes possible and what proceeds from them. When an idea, like robot rights, is immediately declared unthinkable, that is an indication of the need for critical philosophy and the difficult but necessary task of confronting and thinking the unthinkable.

Second, challenging exclusions and prohibitions is the proper work of ethics. Ethics operates by making exclusive decisions. It inevitably and unavoidably picks winners and losers, and determines who is inside the moral community and what remains outside or on the periphery (Gunkel 2014 and Coeckelbergh 2012). But our moral theories and practices develop and evolve by challenging these exceptions and limitations. Ethics, therefore, advances by critically questioning its own exclusivity and eventually accommodating many previously excluded or marginalized others—women, people of color, animals, the environment, etc. "In the history of the United States," Susan Leigh Anderson (2008, 480) has argued, "gradually more and more beings have been granted the same rights that others possessed and we've become a more ethical society as a result." But as Stone (1974, 6) has pointed out, this moral progress is difficult insofar as any extension of rights—any effort to extend moral consideration to previously excluded populations—has always been "a bit unthinkable." In other words, "each time there is a movement to confer rights onto some new 'entity,' the proposal is bound to sound odd or frightening or laughable" (Stone 1974, 7). The task of both moral and legal philosophy, therefore, is not (and cannot be) simply to defend existing orthodoxies by shrinking away in fright in the face of alternatives, triggering angry reactions to the possibility of including others, or mocking such challenges with dismissive laughter. Instead the task is, or at least should be, to stress-test and question the limitations and exclusions of existing moral positions and modes of thinking. Defending orthodoxy is the purview of religion and ideology; critically testing hypotheses and remaining open to revising the way we think about the world in the face of new challenges and opportunities is the task of science. The project of philosophy—and moral philosophy in particular—has always been to ask about what has not been thought before or what has been decried and avoided as unthinkable. Doing so for the previously unthinkable thought of "robot rights" is precisely the objective of the chapters that follow.

## 2 !S1→!S2: Robots Cannot Have Rights; Robots Should Not Have Rights

With the first modality, one infers negation of S2 from the negation of S1. Robots are incapable of having rights (or robots are not the kind of entity that is capable of being a holder of privileges, claims, powers, and/or immunities); therefore robots should not have rights. The assertion sounds intuitive and correct, precisely because it is based on what appears to be irrefutable ontological facts: robots are just technological artifacts that we design, manufacture, and use. A robot, no matter how sophisticated its design and operations, is like any other artifact (e.g., a toaster, a television, a refrigerator, an automobile, etc.); it does not have any particular claim to independent moral or legal status, and we are not and should not be obligated to it for any reason whatsoever. As Johannes Marx and Christine Tiefensee (2015, 83) accurately characterize it:

> Robots are nothing more than machines, or tools, that were designed to fulfill a specific function. These machines have no interests or desires; they do not make choices or pursue life plans; they do not interpret, interact with and learn about the world. Rather than engaging in autonomous decision-making on the basis of self-developed objectives and interpretations of their surroundings, all they do is execute a preinstalled programme. In short, robots are inanimate automatons, not autonomous agents. As such, they are not even the kind of object which could have a moral status.

### 2.1 Default Understanding

This seemingly correct way of thinking is structured and informed by the default answer that is typically provided for the question concerning technology. "We ask the question concerning technology," Martin Heidegger (1977, 4–5) writes, "when we ask what it is. Everyone knows the two statements that answer our question. One says: Technology is a means to an end. The other says: Technology is a human activity. The two definitions of technology belong together. For to posit ends and procure and utilize the means to them is a human activity. The manufacture and utilization of equipment, tools, and machines, the manufactured and used things themselves, and the needs and ends that they serve, all belong to what technology is." According to Heidegger's analysis, the presumed role and function of any kind of technology—whether it be a simple hand tool, a kitchen appliance like a toaster, a jet airliner, or a robot—is that it is a means employed by human users for specific ends. Heidegger terms this particular characterization of technology "the instrumental definition," and he indicates that it forms what is considered to be the "correct" understanding of any kind of technological contrivance.

As Andrew Feenberg (1991, 5) summarizes it: "The instrumentalist theory offers the most widely accepted view of technology. It is based on the common sense idea that technologies are 'tools' standing ready to serve the purposes of users." And because a tool or instrument "is deemed 'neutral,' without valuative content of its own" (Feenberg 1991, 5), a technological artifact is evaluated not in and of itself, but on the basis of the particular employments that have been decided by its human designer or user. Consequently, technology is only a means to an end; it is not and does not have an end in its own right. "Technical devices," as Jean-François Lyotard (1984, 33) writes, "originated as prosthetic aids for the human organs or as physiological systems whose function it is to receive data or condition the context. They follow a principle, and it is the principle of optimal performance: maximizing output (the information or modification obtained) and minimizing input (the energy expended in the process). Technology is therefore a game pertaining not to the true, the just, or the beautiful, etc., but to efficiency: a technical 'move' is 'good' when it does better and/or expends less energy than another." Lyotard's explanation begins by affirming the traditional understanding of technology as an instrument, prosthesis, or extension of human faculties. Given this "fact," which is stated as if it were something that is beyond question, he proceeds to provide an explanation of the proper place of the technological apparatus in epistemology, ethics, and aesthetics. According to his analysis, a technological device, whether it be a corkscrew, a toaster, or a computer, does not in and of itself participate in the big questions of truth, justice, or beauty. Technology is simply and indisputably about efficiency. A particular technological innovation is considered "good," if and only if it proves to be a more effective means to accomplishing a desired end. And this decision is echoed by other critical thinkers of technology: "The moral value of purely mechanical objects," as David F. Channell (1991, 138) explains, "is determined by factors that are external to them—in effect, by the usefulness to human beings." Or as John Searle (1997, 190) explains concerning the computer: "I believe that the philosophical importance of computers, as is typical with any new technology, is grossly exaggerated. The computer is a useful tool, nothing more nor less."

The instrumental theory not only sounds reasonable, it is obviously useful. It is, one might say, instrumental for making sense of things in an age of increasingly complex technological systems and devices. And the theory applies not only to simple devices like corkscrews, toothbrushes, and garden hoses, but also sophisticated technologies, like computers, artificial intelligence, and robots. "Computer systems," Deborah Johnson (2006, 197) asserts, "are produced, distributed, and used by people engaged in social practices and meaningful pursuits. This is as true of current computer systems as it will be of future computer systems. No matter how independently, automatic, and interactive computer systems of the future behave, they will be the products (direct or indirect) of human behavior, human social institutions, and human decision." According to this way of thinking, technologies, no matter how sophisticated, interactive, or seemingly social they appear to be, are just tools; nothing more. They are not—not now, not ever—capable of becoming moral subjects in their own right, and we should not treat them as such. It is precisely for this reason that, as J. Storrs Hall (2001, 2) points out, "we have never considered ourselves to have moral duties to our machines," and that, as David Levy (2005, 393) concludes, the very "notion of robots having rights is unthinkable."

## 2.2 Literally Instrumental

Because the instrumental theory possesses a kind of default status, different forms and configurations of it have been articulated and mobilized in the literature on robots and robotics. This occurs not only in the serious efforts of engineering and scientific research but also in works of fiction. Čapek's *R.U.R.* has been instrumental in this regard, introducing a template for many of the contemporary debates and discussions about these matters. The play's opening dramatic conflict pits the knowledge and experience of the technical staff—the engineers and scientists who manufacture the robotic appliances—against a seemingly naïve outsider, Helena Glory, who comes to the manufacturing facility as a representative of the League of Humanity, a human rights organization that seeks to liberate the robots.

**Fabry (Technical Director):** Can I ask you, what actually is it that your League … League of Humanity stands for?
**Helena:** It's meant to … actually it's meant to protect the robots and make sure … make sure that they're treated properly.
**Fabry:** That is not at all a bad objective. A machine should always be treated properly. In fact I agree with you completely. I never like it when things are damaged. Miss Glory, would you mind enrolling all of us as new paying members of your organization?
**Helena:** No, you do not understand. We want, what we actually want, is to set the robots free!
**Hallemeier (Head of the Institute for Robot Psychology and Behavior):** To do what?
**Helena:** They should be treated … treated the same as people.
**Hallemeier:** Aha. So you mean that they should have the vote! Do you think they should be paid a wage as well?
**Helena:** Well of course they should!
**Hellemeier:** We'll have to see about that. And what do you think they'd do with their wages?
**Helena:** They'd buy … buy things they need … things to bring them pleasure.
**Hallemeier:** This all sounds very nice; only robots do not feel pleasure. And what are these things they're supposed to buy? They can be fed on pineapples, straw, anything you like; it's all the same to them, they haven't got a sense of taste. There's nothing they're interested in, Miss Glory. It's not as if anyone has ever seen a robot laugh.
**Helena:** Why … why … why don't you make them happier?
**Hallemeier:** We couldn't do that, they're only robots after all. They've got no will of their own. No passion. No hopes. No soul. (Čapek 2009, 27–28)

In this early scene from act one of the play, the dramatic conflict concerns, on the one side, authorized experts in robotics, namely the corporation's chief engineer and lead behavioral scientist, who presumably know the technology inside and out, and are able, therefore, to speak about its operations and capabilities with considerable insight and authority. On the other side, there is a naïve and sympathetic "girl," who believes that the robots are people and should be treated accordingly. As a representative of the League of Humanity, Helena Glory wants to free the robots, because they appear to be just like human beings with (as she assumes) similar feelings, desires, and interests. She therefore journeys to the R.U.R. facility to liberate the robots and advocate for the protection of their rights. In this initial scene, her apparently misguided presumptions are "corrected" by the engineer and scientist who explain—based on their knowledge of and experience with the technology—that the robots, despite the fact that they might look like human persons, are nothing more than appliances or technological instruments. As such, the robots, they confidently assure Miss Glory, need nothing, desire nothing, and deserve nothing. This debate operationalizes and turns on a number of important philosophical concepts.

### 2.2.1 Being vs. Appearance

What the robots are takes precedence over and trumps how they appear to be. Miss Glory (as she is called throughout the play) is portrayed as being "mistaken" in her opinions, because she has drawn conclusions about what robots are from how they appear to be. The chief engineer and lead scientist at R.U.R. are able to point out and correct the mistake, because they know not only how the robots are designed to appear (i.e., to look and act in ways that simulate human appearances and behaviors), but, more importantly, what they really are (i.e., mere instruments or appliances in the service of human users and organizations, bereft of personal interests, needs, or desires). All of this is based on and formulated in terms of a profound philosophical distinction that goes all the way back to Plato's "Allegory of the Cave," a kind of parable (or what contemporary philosophers typically call "a thought experiment"), that is recounted by Socrates at the beginning of book VII of the *Republic*.

The allegory concerns an underground cavern inhabited by men who are confined to sit before a large wall on which are projected shadow images. The cave dwellers are, according to the Socratic account, chained in place from childhood and are unable to see anything other than these artificial projections. Consequently, they operate as if everything that appears before them on the wall is, in fact, real and true. They bestow names on the different shadows, devise clever methods to predict their sequence and behavior, and hand out awards to each other for demonstrated proficiency in knowing such things (Plato 1987, 515a–b). At a crucial turning point in the narrative, one of the captives is released. He is unbound by some ambiguous but external action, dragged kicking and screaming out of the cave, and forced to confront the real world that exists outside the subterranean cavern. In beholding the real things in the light of day, the former inmate comes to realize that what he had once thought to be real was just a deceptive shadow and not really what it had appeared to be. And now, armed with this new knowledge, he goes back into the cave in order to correct the mistaken opinion of his colleagues, endeavoring to inform them about what he had learned, namely, that what appears on the wall is not what is really the case.

A more contemporary and AI specific version of this narrative has been developed by John Searle with his Chinese Room thought experiment. This intriguing and rather influential illustration, introduced in 1980 with the essay "Minds, Brains, and Programs" and elaborated in subsequent publications, was offered as an argument against the claims of strong AI—that machines are able to achieve actual thought:

> Imagine a native English speaker who knows no Chinese locked in a room full of boxes of Chinese symbols (a data base) together with a book of instructions for manipulating the symbols (the program). Imagine that people outside the room send in other Chinese symbols which, unknown to the person in the room, are questions in Chinese (the input). And imagine that by following the instructions in the program the man in the room is able to pass out Chinese symbols which are correct answers to the questions (the output). The program enables the person in the room to pass the Turing Test for understanding Chinese but he does not understand a word of Chinese. (Searle 1999, 115)

The point of Searle's imaginative, albeit somewhat ethnocentric, illustration ("ethnocentric" insofar as Chinese has been European philosophy's exotic "other" at least since the time of Leibniz, see Perkins 2004) is quite simple—appearance is not the real thing. Merely shifting symbols around and exchanging one for another in a way that looks like linguistic understanding is not actually an understanding of language. Demonstrations like Searle's Chinese Room, which differentiate between how something appears to be and what that thing actually is, not only mobilize the ancient philosophical distinction inherited from Plato, but inevitably require, as the Platonic allegory had already exhibited, some kind of privileged and immediate access to the true nature of things and not just how they appear to be. Although there remains considerable debate within philosophy about this matter, specifically the realism/antirealism debate, it is often not a difficulty in the empirical sciences and the practical efforts of engineering, where one can, as illustrated in *R.U.R.*, simply differentiate between the authority of knowledgeable experts, who know the true nature of things, and the naïve opinions and beliefs of uniformed outsiders, who are often limited to how things appear to be.

### 2.2.2  Ontology Precedes Ethics

In confronting and dealing with other entities—whether other human persons, mere artifacts like a toaster, or the human-looking robots depicted in *R.U.R.*—one inevitably needs to distinguish between those beings *who* are in fact moral subjects and *what* remains a mere thing. As Jacques Derrida (2005, 80) explains, the difference between these two small and seemingly insignificant words—"who" and "what"—makes a big difference, precisely because it parses the world of entities into two camps: those Others who can and should have a legitimate right to privileges, claims, powers and/or immunities and mere things that are and remain objects, instruments, or artifacts.[1] These decisions (which are quite literally a cut or *de-caedere* in the fabric of being) are often accomplished and justified on the basis of intrinsic, ontological properties. "The standard approach to the justification of moral status is," Mark Coeckelbergh (2012, 13) explains, "to refer to one or more (intrinsic) properties of the entity in question, such as consciousness or the ability to suffer. If the entity has this property, this then warrants giving the entity a certain moral status." In this transaction, ontology always takes precedence over ethics. What something is determines how it ought to be treated. Or as Luciano Floridi (2013, 116) describes it: "what the entity is determines the degree of moral value it enjoys, if any." According to this standard procedure, the question concerning the status of others—whether they are some*one who* matters or some*thing* that does not—would need to be resolved by first identifying which property or properties would be necessary and sufficient for moral status, and then figuring out whether a particular entity possesses that property or not. As Hume already knew and criticized, decisions concerning how something ought to be treated are usually derived from what it is.

The problem (or "the opportunity," since it all depends on how one looks at it) is that the decision between *who* and *what* has never been static or entirely settled; it is continuously challenged and open to considerable renegotiation. The history of moral philosophy can, in fact, be read as something of an ongoing debate and struggle over where one makes the cut or draws the line. Ethics has, therefore, evolved in fits and starts as the product of different "liberation movements," as Peter Singer (1989, 148) calls it, whereby what had been previously considered a mere thing comes to be recognized, for one reason or another, as a legitimate subject of moral concern, passing from the side of *what* to the side of *who*. Despite these progressive developments whereby many things that had been regarded as mere things—at one time people of color, women, children, animals, etc.—come to be recognized as legitimate moral subjects, mechanisms like robots appear to be perennially stuck on the side of *what*. "Human history," as Inayatullah and McNally (1988, 123) point out: "is the history of exclusion and power. Humans have defined numerous groups as less than human: slaves, woman, the 'other races,' children and foreigners. These are the wretched who have been defined as, stateless, personless, as suspect, as rightless. This is the present realm of robotic rights." For this reason, and as I have previously argued (Gunkel 2012 and 2017b), machines in general, and robots in particular, have been (and apparently continue to be) the one thing that remains excluded from moral philosophy's own efforts to achieve greater levels of inclusion. No matter how one comes to decide or to renegotiate the difference between *who* counts and *what* does not, robots are not moral subjects; they are and will remain mere things—an instrument or "means to an end," but not an end in their own right.[2] And this difference—the difference between *what* is a mere thing and *who* is another legitimate subject—is not only deployed and operationalized in *R.U.R.* but is an organizing principle throughout robot science fiction, including *Bladerunner*, *Westworld*, and *Battlestar Galactica*:

**Admiral Adama:** She was a Cylon, a machine. Is that what Boomer was, a machine? A thing?
**Chief Tyrol:** That's what she turned out to be.
**Admiral Adama:** She was more than that to us. She was more than that to me. She was a vital, living person aboard my ship for almost two years. She couldn't have been just a machine. Could you love a machine? (*Battlestar Galactica*, 2005)

### 2.2.3 Limited Rights

The robots of *R.U.R.* are introduced and characterized as mere mechanisms that are designed and employed by human beings to serve the purposes and interests of human society. They are, therefore, determined to be instruments, possessing no independent moral or legal status of their own. As such, and by definition, they neither can nor should they have rights. Or perhaps stated more accurately, whatever "rights"—whatever privileges, claims, powers, or immunities—they might enjoy or deserve are limited to what would be granted to any other piece of valuable property. This is, in fact, the root cause of the initial misunderstanding between Fabry, the engineer, and Helena Glory. For Miss Glory, "protection" means nothing less than freeing the robots from the burden of their labor. She assumes that the robots, because they look and appear to act like people, must also possess similar interests, and that, because of this, they have a claim-right to be protected from burdensome labor and abuse. For the engineer, however, "protection" has an entirely different denotation. On his account—or at his particular "level of abstraction," to mobilize terminology developed by Floridi (2013)—sophisticated instruments, like the robots, are valuable artifacts and assets. They should, therefore, be treated properly to ensure their continued operation and to prevent damage to the instrument that would render it unusable. For this reason, the engineer's position concerning the relative status of the robot is similar to that expressed by Laszlo Versenyi (1974, 252):

Questions such as whether robots should be blamed or praised, loved or hated, given rights and duties, etc., are in principle the same sort of questions as, "Should cars be serviced, cared for, and supplied with what they require for their operation?" To the extent that we depend on cars and servicing them is necessary for their well-functioning, it is as prudent not to withhold such services from them as it is not to withhold comparable, operationally required services from anything else on whose well-functioning we depend. In either case the only relevant considerations are: Do we need these things (men, animals, machines) for our well-functioning, and do they need this or that (care, love, food, gas, control) for theirs? As soon as both questions are answered there are no further considerations relevant to our decision.

A similar argument is recognized by Coeckelbergh. Even "if we regard robots as things," Coeckelbergh (2010b, 240) writes, we may still have "some obligations to treat them well in so far as they are the property of humans or in so far as they have value for us in other ways." Even if the robot cannot itself have rights (presumably because it is a mere thing and therefore beyond good and evil), there are reasons that we should treat it well. "The rationale to respect the robot here is not that the robot has moral agency or moral patiency, but instead that it belongs to a human and has value for that human person and that in order to respect that other human person we have certain indirect obligations towards robots-as-property. ...

Since things are valuable to us (humans) for various reasons, it is likely that robots will receive some degree of indirect moral consideration anyway" (Coeckelbergh 2010b, 240). In other words, just because something is a thing and not capable of having rights (i.e., privileges, claims, powers, and/or immunities) in its own right that certainly does not mean that we do not have obligations in the face of such things. "We do then," as Joanna Bryson (2010, 73) recognizes, "have obligations regarding robots, but not really to them." Consequently to say that robots are bereft of some basic protections is not entirely accurate. Robots like any other artifact are due some consideration necessary to maintain their instrumental utility. But this respect for the object is not anything like the rights that are assumed to be in play by *R.U.R.'s* Helena Glory and the League of Humanity.

## 2.3   Instrumentalism at Work

In *R.U.R.*, the robots are introduced and characterized as mere mechanisms that do not have interests and therefore are incapable of having rights or of being liberated. From the informed perspective of the engineer and scientist, Helena Glory's (misguided) plan to free the robots is not just unnecessary, it is pure nonsense. It would be as absurd as seeking to liberate your toaster from the burdens of making toast so that it could pursue its own interests and desires.[3] Similar decisions and arguments have been instituted and mobilized in the scientific literature. One of the more influential and contemporary articulations can be found in the "Principles of Robotics" (Boden et al. 2011 and 2017). The history of the document and the manner of its composition is not immaterial. "In 2010," Tony Prescott and Michael Szollosy (2017, 119) explain: "the UK's Engineering and Physical Science and Arts and Humanities Research Councils (EPSRC and AHRC) organised a retreat to consider ethical issues in robotics to which they invited a pool of experts drawn from the worlds of technology, industry, the arts, law and social sciences. This meeting resulted in a set of ethical 'Principles of Robotics' (henceforth 'the Principles') that were published online by the EPSRC (Boden et al., 2011), which aimed at 'regulating robots in the real world.'"

The "pool of experts" included some of the most noted and recognized names in the field of AI and robotics—Margaret Boden, Joanna Bryson, Darwin Caldwell, Kerstin Dautenhahn, Lilian Edwards, Sarah Kember, Paul

Newman, Vivienne Parry, Geoff Pegman, Tom Rodden, Tom Sorrell, Mick Wallis, Blay Whitby, and Alan Winfield—and the document they produced consisted of five rules, "presented in a semi-legal version together with a looser, but easier to express, version that captures the sense for a non-specialist audience" (Boden et al. 2017, 125) and "seven high-level messages ... designed to encourage responsibility within the robotics research and industrial community" (Boden et al. 2017, 128). The Principles were subsequently re-visited and re-evaluated during a 2016 AISB (Society for Study of Artificial Intelligence and Simulation of Behavior) workshop convened in Sheffield, UK. The workshop was chaired by Tony Prescott, and the organizing committee included Joanna Bryson and Alan Winfield, who had contributed to the original document, along with Madeleine de Cock Buning and Noel Sharkey. The revised "Principles" along with the fourteen responses and commentaries that were initially developed for the workshop were then published in a special issue of *Connection Science* in 2017. Although the document is rather short, its composition and content have a number of important consequences.

### 2.3.1 Expertise

The principles were initially devised, critically revised, and finally approved by noted experts in the field. The authorship of "The Principles" is not insignificant, and the framing of the document makes it absolutely clear who has the authority to speak on this matter: "In September 2010, experts drawn from the worlds of technology, industry, the arts, law and social sciences met at the joint EPSRC and AHRC Robotics Retreat to discuss robotics, its applications in the real world and the huge amount of promise it offers to benefit society"[4] (Boden et al. 2011, 1). This claim on "expertise" is significant. The authors of the document are informed insiders who presumably know about robots, i.e., what these things are and are capable of doing or not doing. Like the engineer and the chief scientist in *R.U.R.*, these experts presumably have the experience, insight, and knowledge to speak with authority about these matters and are authorized to correct potential misunderstandings like those expressed by Helena Glory and her League of Humanity. The fact that "The Principles" have been authored by a group of experts is part of its rhetorical strategy. In the case of this document, it definitely "makes a difference who the speaker is and where s/he comes from" (Plato 1982, 275b).

## 2.3.2 Robots Are Tools

According to the experts, robots are tools and cannot and should not be considered persons—entities who have moral and legal standing and possess rights and/or responsibilities. This is articulated in various ways and numerous times throughout the course of the document:

- "Robots are simply tools of various kinds, albeit very special tools, and the responsibility of making sure they behave well must always lie with human beings" (Introduction).
- "Robots are multi-use tools" (Rule #1).
- "Robots are just tools, designed to achieve goals and desires that humans specify" (Commentary to Rule #2).
- "Robots are products" (Rule #3).
- "Robots are manufactured artefacts" (Rule #4).
- "Robots are simply not people" (Commentary to Rule #3) (Boden et al. 2017).

Even if or when robots are designed to simulate something more than a mere tool, e.g., a pet or a companion, this appearance, it is argued, should be clearly indicated as such and the actual "tool-being" (to use Graham Harman's [2002] Heideggerian influenced terminology) of the robot should be made easily accessible and readily apparent:

> One of the great promises of robotics is that robot toys may give pleasure, comfort and even a form of companionship to people who are not able to care for pets, whether due to restrictions in their homes, physical capacity, time or money. However, once a user becomes attached to such a toy, it would be possible for manufacturers to claim the robot has needs or desires that could unfairly cost the owners or their families more money. The legal version of this rule was designed to say that although it is permissible and even sometimes desirable for a robot to sometimes give the impression of real intelligence, anyone who owns or interacts with a robot should be able to find out what it really is and perhaps what it was really manufactured to do. Robot intelligence is artificial, and we thought that the best way to protect consumers was to remind them of that by guaranteeing a way for them to "lift the curtain" (to use the metaphor from The Wizard of Oz). (Boden et al. 2017, 127)

No matter how sophisticated their behavior or elegant their design, "The Principles" make it clear that robots are instruments or tools to be used, deployed, and manipulated by human agents. Consequently, the document endorses and operationalizes what Heidegger (1977, 6) identified as the "instrumental and anthropological definition of technology."

"The ESPRC Principles," as Szollosy (2017, 151) concludes, "make certain, very specific, yet completely unspoken assumptions as to what constitutes a 'human being.' And the Principles suit thus [SIC] human being very well. The Principles insist upon a particular relationship between clearly demarcated human subjects, on the one hand, who always act as unique and autonomous agents, and robots, which are forever seen as objects, tools to be manipulated by human masters."

### 2.3.3  Is/Ought Inference

In "The Principles," *ought* is derived from *is*. How something should be treated is entirely determined and justified by what ontological properties it has or is capable of possessing. As Prescott explains in his commentary "Robots are not just Tools":

> At the heart of the EPSRC principles of robotics (henceforth "the principles") are a number of ontological claims about the nature of robots that serve as axioms to frame the subsequent development of ethical challenges and rules. These include claims about what robots are, and also about what they are not. The claims about what robots are include that "robots are multiuse tools" (principle 1), that "robots are products" (principle 3) and "pieces of technology" (commentary on principle 3), and that "robots are manufactured art[i]facts" (principle 4). The claims about what robots are not include that "humans, not robots, are responsible agents" (principle 2), that robots are "simply not people" (commentary on principle 3), and that robot intelligence can give only an "impression of real intelligence" (commentary on principle 4). (Prescott 2017, 142)

As Prescott points out, "The Principles" proceed by first issuing and operationalizing a number of assertions about the ontological conditions and status of robots. These directly stated claims then function as axioms—statements or propositions that are regarded as being established, accepted, or self-evidently true—from which are derived a set of ethical rules and guidelines. Determinations about how robots ought to be regarded, therefore, proceed from what robots are assumed to be or not be. And since robots are simply declared to be just tools and not persons, it then follows that they do not have responsibilities or any claim to rights. In this argument, as Prescott insightfully points out, everything proceeds from and depends on the initial ontological decree and assumption; *ought* is derived from *is*.

## 2.4 Duty Now and for the Future

"The Principles" were designed for, and are intentionally and quite consciously focused on, current opportunities and challenges. But things could change. "One of the consequences of the view of robots as 'just tools,'" as Prescott (2017, 147) points out in his dissenting opinion on this matter, "is the implicit dismissal of the possibility of strong AI—that future robots could have human-level, or beyond human-level general intelligence." But even when considered from this future-oriented (and rather speculative) perspective, there may still be reasonable arguments for excluding robots from the community of moral subjects. Consider, for instance, what Lantz Fleming Miller (2015, 374) calls "maximally humanlike automata." "My concern," Miller writes, "is whether automata that exhibit all (or sufficiently close to all) traits considered to be distinctive and necessary for being a human should thereby enjoy full human rights" (Miller 2015, 375). This concern is an extreme case that is, at least from our current vantage point, entirely speculative: If it were possible to construct automata that exhibit all the traits that are necessary for an entity to be considered human—"all except being biologically human" (Miller 2015, 380)—should these artifacts then have similar, if not the same, rights (e.g., privileges, claims, powers, and/or immunities) as that enjoyed by a human person? Miller answers this question with a resounding and unqualified "no." In other words, even when pushed to the extreme limit of possibility, "one is not obliged to grant humanlike automata full human rights" (Miller 2015, 377).

The reason for this, as Miller argues, is that human beings and automata are ontologically distinct. As biological entities that are the product of evolution, human beings exhibit "existential normative neutrality," while automata, which are artifacts created by human beings, do not. For this reason, human beings can be defined by way of a "one-place predicate $A(x)$," which is to say that "X has come into existence ('arisen'/'come to be')," whereas artifacts must be defined by way of a two or even three place predicate: $C(x,y)$ "some entity Y has constructed X" and $P(x,y,z)$ "Some entity Y has constructed X for purpose Z" (Miller 2015, 375). Although he is no Heideggerian by any stretch of the imagination, Miller's formulation is remarkably close to Heidegger's analysis of instrumentality in *Being and Time* (1962, 102) even if the vocabulary is a bit different: "Intrinsic properties are considered to be necessary for the object to be the kind it is. An extrinsic property is only incidental or contingent to the object. For artifacts, the purpose is thereby an intrinsic property. A shovel is defined as an object made to dig. A computer is made to calculate. Without the defined purpose, the object is a mass of metal, not a shovel" (Miller 2015, 377). In order for a shovel to be a shovel, it has a purpose for which it is made and to which it is applied. It is this purpose, this "for which" (as Heidegger would say), that makes the shovel a shovel and not just an obtrusive chunk of steel attached to a long shaft of wood. Although the vocabulary is different—most likely owing to differences in the continental and analytic traditions of philosophy—the characterization is virtually identical. Things like robots are what they are insofar as they have a teleology, a purpose to which they are applied and already destined. Human beings, by contrast, lack this teleological orientation; they just are.[5]

In making this argument, Miller follows and endorses that longstanding tradition in moral philosophy where ontology precedes ethics such that one derives a determination of *ought* from *is*. According to Miller, it is because human beings (who are the product of evolution and therefore exhibit what he calls "existential normative neutrality") and humanlike automata (which are things or instruments that are intentionally designed with a teleological purpose) are ontologically different from one another that the former is deserving of rights and the latter can be defensibly denied the same. And since this particular brand of human exceptionalism is grounded in what Miller argues is a fundamental and incontrovertible intrinsic difference, it is (or Miller at least asserts that it is) absolutely fair and guilt free:

> Humans are the ones who discern, affirm, and thereby realize human rights. It is fair and just that they grant such rights, and they occupy a fair and just position to determine which ontological kinds of entities deserve recognition of rights. Particularly, humans are under no moral obligation to grant full human rights to entities possessing ontological properties critically different from them in terms of human rights bases. To grant full human rights recognition only to *Homo sapiens* does not run against the basis of human rights. Placing the rights partition between humans and automata does not hark back to eras when human rights were granted only to white, European males. We need not attempt to circumvent the return to those eras by extending rights to all kinds of ontologically varying entities, just in case they exhibit some traits that humans have. (Miller 2015, 387)

Whether this "ontological distinction" is indeed a philosophically defensible position is something that remains debated. In fact, as Miller (2015, 374) explicitly notes, the entire argument is "hypothetical, resting conditionally upon a widely held belief about the nature of human rights and the properties of human beings." Miller's argument, therefore, is ultimately grounded not in a scientifically proven or even provable fact but in a common, "widely held belief"; it is a matter of faith. Szollosy makes a similar point concerning "The Principles":

> At the (implicit) heart of the ESPRC Principles is a particular human being defined through the last centuries by what has becomes known as *humanism*. This human being is an agent in its own right, a being that is independent and not to be governed by other, metaphysical, or supernatural, forces. This human being is at the centre of European-based legal, ethical, economic and political systems; however, it is vital to remember that 1) this human being is still a relatively new invention and that 2) throughout its life-span, there has never just been one, singular version of this human being, as humanist proponents have liked to imagine that it is. (Szollosy 2017, 151)

When one derives *ought* from *is* (or *ought not* from *is not*), ontology determines everything. But decisions about ontological status tend to rest on dogmatic assertions that are contingent and therefore open to significant critical challenges.

What Miller calls a "widely held belief," for instance, can be easily challenged by other beliefs that are just a valid (or invalid, as the case may be). Nick Bostrom and Eliezer Yudkowsky (2014) propose an equally plausible alternative by way of something they call non-discrimination principles (which as "principles" are also not argued but dogmatically asserted as if they were unquestionably true):

> Principle of Substrate Non-Discrimination—If two beings have the same functionality and the same conscious experience, and differ only in the substrate of their implementation, then they have the same moral status.
>
> Principle of Ontogeny Non-Discrimination—If two beings have the same functionality and the same consciousness experience, and differ only in how they came into existence, then they have the same moral status. (Bostrom and Yudkowsky 2014, 322–23)

These two principles, which as principles are not established through argument or evidence but simply asserted, derive a decision concerning moral status not from how something is—its ontological distinctiveness—but from how it appears to functions in actual experience.

A similar argument is proposed by Eric Schwitzgebel and Mara Garza (2015, 100): "It shouldn't matter to one's moral status what kind of body one has, except insofar as one's body influences one's psychological and social properties. Similarly, it shouldn't matter to one's moral status what kind of underlying architecture one has, except insofar as underlying architecture influences one's psychological and social properties. Only psychological and social properties are directly relevant to moral status—or so we propose." For this reason, proposals like that offered by Miller are declared to be both atypical and archaic:

> All of the well-known modern secular accounts of moral status in philosophy ground moral status only in psychological and social properties, such as capacity for rational thought, pleasure, pain, and social relationships. No influential modern secular account is plausibly read as committed to a principle whereby two beings can differ in moral status but not in any psychological or social properties, past, present, or future, actual or counterfactual. However, some older or religious accounts might have resources to ground a difference in moral status outside the psychological and social. An Aristotelian *might* suggest that AIs would have a different *telos* or defining purpose than human beings. However, it's not clear that an Aristotelian must think this; nor do we think such a principle, interpreted in such a way, would be very attractive from a modern perspective, unless directly relevant psychological or social differences accompanied the difference in telos. Similarly, a theist *might* suggest that God somehow imbues human beings with higher moral status than AIs, even if they are psychologically and socially identical. We find this claim difficult to assess, but we're inclined to think that a deity who distributed moral status unequally in this way would be morally deficient. (Schwitzgebel and Garza 2015, 100)

No matter how it is formulated, however, these differing accounts make the same basic argumentative gesture—they derive moral status from rather dogmatic assertions/assumptions about "what is the case." Ontological assumptions, in other words, determine and justify moral consideration, which had been the problem Hume initially identified and critiqued by way of the is/ought distinction.

## 2.5 Complications, Difficulties, and Potential Problems

Although this way of thinking works and has been instrumental for resolving questions of who has moral status and what does not, there are a number of significant critical problems.

### 2.5.1 Tool != Machine

The instrumental theory, for all its success in helping us make sense of technological innovation, is a rather blunt instrument, reducing all technology, irrespective of design, construction, or operation, to a tool or instrument. "Tool," however, does not necessarily encompass everything technological, and does not, therefore, exhaust all possibilities. There are also *machines*. Although "experts in mechanics," as Karl Marx (1977, 493) pointed out, often confuse these two concepts calling "tools simple machines and machines complex tools," there is an important and crucial difference between the two. Indication of this essential difference can be found in a brief parenthetical remark offered by Heidegger in "The Question Concerning Technology." "Here it would be appropriate," Heidegger (1977, 17) writes in reference to his use of the word "machine" to characterize a jet airliner, "to discuss Hegel's definition of the machine as autonomous tool [*selbständigen Werkzeug*]." What Heidegger references, without supplying the full citation, are Hegel's 1805–7 Jena Lectures, in which "machine" had been defined as a tool that is self-sufficient, self-reliant, or independent. As Marx (1977, 495) succinctly described it, picking up on this line of thinking, "the machine is a mechanism that, after being set in motion, performs with its tools the same operations as the worker formerly did with similar tools."

Understood in this way, Marx (following Hegel) differentiates between the tool used by the worker and the machine, which does not occupy the place of the worker's tool but takes the place of the worker him/herself. Although Marx did not pursue an investigation of the social, legal, or moral consequences of this insight, it produces some interesting ambiguities with the assignment of moral and legal accountability. As John Sullins (2005, 1) explains, "since an autonomous machine is a surrogate actor for a human agent that would have accomplished the tasks of the machine in its absence, then it follows that the machine is also a surrogate for the ethical rights and responsibilities that apply to a human actor in the same situation." Because the autonomous machine replaces not the tool but the human user of the tool, the machine can be considered a surrogate for the rights and responsibilities that had been assigned to the human agent. And this seems to have traction with recent developments that have advanced explicit proposals for robots—or at least certain kinds of robots—to be defined as something other than mere instruments or tools.

Perhaps the best illustration of the difference Marx identifies and describes is available with the self-driving car or autonomous vehicle. The autonomous vehicle, whether the Google Car or one of its competitors, is not designed for and intended to replace the automobile. It is, in its design, function, and materials, the same kind of instrument that we currently utilize for the purpose of personal transportation. The autonomous vehicle, therefore, does not replace the instrument of transportation (the automobile); it is intended to replace (or at least significantly displace) the driver. This difference was acknowledged by the US National Highway Traffic Safety Administration (NHTSA), which in a February 4, 2016 letter to Google, stated that the company's Self Driving System (SDS) could legitimately be considered the legal driver of the vehicle: "As a foundational starting point for the interpretations below, NHTSA will interpret 'driver' in the context of Google's described motor vehicle design as referring to the SDS, and not to any of the vehicle occupants" (Ross 2016). Although this decision is only an interpretation of existing law, the NHTSA explicitly states that it will "consider initiating rulemaking to address whether the definition of 'driver' in Section 571.3 [of the current US Federal statute, 49 U.S.C. Chapter 301] should be updated in response to changing circumstances" (Hemmersbaugh 2016).

Similar proposals have been floated in efforts to deal with workplace automation. In a highly publicized draft document submitted to the European Parliament in May of 2016, for instance, it was argued that "sophisticated autonomous robots" ("machines" in Marx's terminology) be considered "electronic persons" with "specific rights and obligations" for the purposes of contending with the challenges of technological unemployment, tax policy, and legal liability. Although the proposal did not pass as originally written, it represents recognition on the part of policymakers that recent innovations in robotics challenge the way we typically respond to and answer questions regarding moral and legal responsibility. In both these cases, we have decisions that recognize the robotic mechanism as being more than just an instrument through which human beings act on the world. Such technologies, as David C. Vladeck (2014, 121) explains, "will not be tools *used* by humans; they will be machines *deployed* by humans that will act independently of direct human instruction, based on information the machine itself acquires and analyzes, and will often make highly consequential decisions in circumstances that may not be anticipated by, let alone directly addressed by, the machine's creators." Such mechanisms, therefore, challenge the instrumental theory and lead to some recognition of the robot as a kind of independent moral and/or legal entity. Whether any of this is eventually codified in policy or law remains to be seen. But what we can see at this point is a challenge to the instrumentalist way of thinking about technology that is opening up other opportunities for recognizing the social position and standing of these mechanisms.

### 2.5.2 Not Just Tools

Additionally, the instrumental theory appears to be unable to contend with recent developments in social robotics. "Tool" might not fit (or at least not easily accommodate) things that have been intentionally designed to be social companions. "The category of tools," as Prescott (2017, 142) explains, "describes physical/mechanical objects that serve a function, whereas the category of companions describes significant others, usually people or animals, with whom you might have a reciprocal relationship marked by emotional bond. The possibility that robots could belong to both these categories raises important and interesting issues that are obscured by insisting that robots are just tools."[6] In fact, practical experiences with socially interactive machines push against the explanatory capabilities of the instrumental theory, if not forcing a break with it altogether. "At first glance," Kate Darling (2016a, 216) writes, "it seems hard to justify differentiating between a social robot, such as a Pleo dinosaur toy, and a household appliance, such as a toaster [again with the toasters]. Both are man-made objects that can be purchased on Amazon and used as we please. Yet there is a difference in how we perceive these two artifacts. While toasters are designed to make toast, social robots are designed to act as our companions."

In support of this claim, Darling offers the work of Sherry Turkle and the experiences of US soldiers in Iraq and Afghanistan. Turkle, who has pursued a combination of observational field research and interviews in clinical studies, identifies a potentially troubling development she calls "the robotic moment": "I find people willing to seriously consider robots not only as pets but as potential friends, confidants, and even romantic partners. We don't seem to care what their artificial intelligences 'know' or 'understand' of the human moments we might 'share' with them ... the performance of connection seems connection enough" (Turkle 2011, 9). In the face of sociable robots, Turkle argues, we seem to be willing, all too willing, to consider these machines to be much more than tools or instruments; we address them as surrogate pets, close friends, personal confidants, and even paramours. Even if it is true, as Matthais Scheutz (2012,

215) asserts, "that none of the social robots available for purchase today (or in the foreseeable future, for that matter) care about humans," that does not seem to matter. Our ability to care for a robot, like our ability to care for an animal, does not seem to be predicated on its ability to care or to show evidence of caring for us. We don't, in other words, seem to care whether they actually care nor not.

But this is not limited to objects like the Furbie, Pleo, AIBO, and PARO robots, which have been intentionally designed to elicit this kind of emotional response. Human beings appear to be able to do it with just about any old mechanism, like the very industrial-looking Explosive Ordnance Disposal (EOD) robots that are being utilized on the battlefield. As Peter W. Singer (2009), Joel Garreau (2007), and Julie Carpenter (2015) have reported, soldiers form surprisingly close personal bonds with their units' EOD robots, giving them names, awarding them battlefield promotions, risking their own lives to protect that of the robot, and even mourning their death. This happens as a product of the way the mechanism is situated within the unit and the role that it plays in battlefield operations. And it happens in direct opposition to what otherwise sounds like good common sense: "They are just technologies—instruments or tools that feel nothing." As Eleanor Sandry (2015a, 340) explains:

> EOD robots, such as PackBots and Talons, are not humanlike or animal-like, are not currently autonomous and do not have distinctive complex behaviours supported by artificial intelligence capabilities. They might therefore be expected to raise few critical issues relating to human-robot interaction, since communication with these machines relies on the direct transmission of information through radio signals, which have no emotional content and are not open to interpretation. Indeed, the fact that these machines are broadly not autonomous precludes them from being discussed as social robots according to some definitions. ... In spite of this, there is an increasing amount of evidence that EOD robots are thought of as team members, and are valued as brave and courageous in the line of duty. It seems that people working with EOD robots, even though the robots are machinelike and under the control of a human, anthropomorphise and/or zoomorphise them, interpreting them as having individual personalities and abilities.

Similar results have been obtained with studies regarding more mundane technological objects, like Guy Hoffmann's AUR desk lamp (Sandry 2015a and 2015b) and the Roomba robotic vacuum clearer (Sung et al. 2007). As Scheutz (2012, 213) reports: "While at first glance it would seem that the Roomba has no social dimension (neither in its design nor in its behavior) that could trigger people's social emotions, it turns out that humans, over time, develop a strong sense of gratitude toward the Roomba for cleaning their home. The mere fact that an autonomous machine keeps working for them day in and day out seems to evoke a sense of, if not urge for, reciprocation."

None of this is necessarily new or surprising. Evidence of it was already tested and demonstrated with Fritz Heider and Mariane Simmel's "An Experimental Study of Apparent Behavior" (1944), which found that human subjects tend to attribute motive and personality to simple animated geometric figures. Similar results have been obtained by way of the computer as social actor (CASA) studies conducted by Byron Reeves and Clifford Nass in the mid-1990s. As Reeves and Nass discovered across numerous trials with human subjects, users (for better or worse) have a strong tendency to treat socially interactive technology, no matter how rudimentary, as if they were other people.

Computers, in the way that they communicate, instruct, and take turns interacting, are close enough to human that they encourage social responses. The encouragement necessary for such a reaction need not be much. As long as there are some behaviors that suggest a social presence, people will respond accordingly. When it comes to being social, people are built to make the conservative error: When in doubt, treat it as human. Consequently, any medium that is close enough will get human treatment, even though people know it's foolish and even though they likely will deny it afterwards. (Reeves and Nass 1996, 22)

So what we have is a situation where our theory of technology—a theory that has considerable history behind it and that has been determined to be as applicable to simple hand tools as it is to complex computer systems—seems to be out of sync with the very real practical experiences we now have with machines in a variety of situations and circumstances. To put it in terminology that is more metaphysically situated: How something appears to be—how it actually operates in real social situations and circumstances—might be more important than what it actually is (or has been assumed to be). "We should," as Prescott (2017, 146) concludes formulating a kind of phenomenological moral maxim, "take into account how people *see* robots, for instance, that they may feel themselves as having meaningful and valuable relationships with robots, or they may see robots as having important internal states, such as the capacity to suffer, despite them not having such capacities."

## 2.5.3 Ethnocentrism

This way of dividing the world into who and what—human persons who matter and mere technological things that do not—is culturally specific and exposed to a kind of moral relativism that is often not adequately identified or accounted for. The ESPRC Principles, for example, have been criticized for assuming a particular concept of "human nature" that is, as Szollosy (2017, 156) argues, "very much a European, Christian concoction."

> Robots are considered to be machines, and therefore merely objects. In the European Christian tradition, such non-living, or even non-human objects, are considered lesser beings on the basis that they do not have a soul; that intangible, metaphysical property unique to life or, in most articulation, unique specifically to humans. (This idea of lacking something vitally human lies at the very idea of the robot, when the word was first introduced to the world in Karel Čapek's 1921 play, *R.U.R.*) Though one could argue that Europe is no longer beholden to Christianity, Europe's (and America's) Christian values are constantly on display, and this assumption is obvious even in contemporary, completely secular European legal and ethical frameworks, including these ESPRC Principles. (Szollosy 2017, 156)

By way of contrast, other religious/philosophical traditions mobilize entirely different ways of thinking about these things. As Jennifer Robertson points out, the instrumental and anthropological viewpoint would be antithetical to a perspective influenced by Japanese culture and traditions. "Three key sociocultural factors," Robertson (2014, 576) argues, "influence the way Japanese experience robots as 'living' entities. The first is linguistic: In Japanese, two separate verbs can be used to describe existence. *Aru/arimasu* refers to the existence of something, a bicycle, for example. *Iru/imasu* is used to refer to the existence of someone. *Iru/imasu* is also used in reference to robots, as in the title *Robotto no iru kurashi* (lit., a lifestyle where robots exist)." The second factor is Shintoism, a religion that arranges things differently than the three dominant monotheisms (Judaism, Christianity, and Islam). "Shinto, the native [Japanese] animistic beliefs about life and death, holds that vital energies, deities, forces, or essences called *kami* are present in both organic and inorganic matter and in naturally occurring and manufactured entities alike. Whether in trees, animals, mountains, or robots, these *kami* (forces) can be mobilized" (Robertson 2014, 576). The third factor, Robertson argues, concerns a different conception of the meaning of life and what is (and what is not) living.

*Inochi*, the Japanese word for "life," encompasses three basic, seemingly contradictory but inter-articulated meanings: a power that infuses sentient beings from generation to generation; a period between birth and death; and, most relevant to robots, the most essential quality of something, whether organic (natural) or manufactured. Thus robots are experienced as "living" things. The important point to remember here is that there is no ontological pressure to make distinctions between organic/inorganic, animate/inanimate, human/nonhuman forms. On the contrary, all of these forms are linked to form a continuous network of beings. (Robertson 2014, 576)

In other words, the instrumental and anthropological definition of technology—that seemingly correct characterization of technology that renders a robot a mere tool as opposed to someone who counts as a socially significant other—will only work when deployed from and within a particular culture and linguistic tradition. It is precisely for this reason, that Veruggio (2005, 4) had called for the robotics community to "develop a general Survey on the main ethical paradigms in the different cultures, religions, faiths" and "define a Rosetta Stone of the ethical guidelines 'adjusted' to the different cultures, religions, faiths."

## 2.6 Summary

A negative reply to the question "Can and should robots have rights?" appears to be simple and intuitive. This default response proceeds from the apparently common-sense understanding that technology—any technology, whether a simple hand tool like a hammer, a household appliance like a toaster (because it is always about toasters), or a sociable robot—is nothing more than a tool or instrument of human activity. This conceptualization is grounded in what Heidegger (1977, 6) calls "the instrumental and anthropological definition" and its seemingly wide acceptance makes it what many consider to be the correct way of thinking—so "correct," in fact, that one does not even need to think about it. This orthodoxy is clearly evident and in play in many attempts to grapple with the moral/legal opportunities and challenges of robots, especially efforts to formulate general principles for the proper integration of the technology in contemporary society. These proposals derive *ought* from *is*, arguing that what robots are (mere inanimate objects and not moral or legal subjects) determine how they should be treated. Although this way of thinking works and appears to be incontrovertible, there are significant limitations and difficulties inherent in it.

First, the ontological category "tool," although applicable to many technologies, does not adequately explain or account for every kind of mechanism, especially what Marx had called machines, which constitute a third kind of liminal entity (Kang 2011, 35) that does not fall neatly on one or the other side of the *who* versus *what* distinction. Second, there are socially situated artifacts that have been deliberately designed to be more than mere tools, like a sociable robot, or that function in ways that make them—in the eyes of their human partners—something more than an instrument. In the face of these other kinds of socially interactive others, how we decide to respond to these entities appears to be more important and significant than what we have been told they actually are. In other words, the social circumstances of the relationship we have with the artifact appear to take precedence over the ontological properties that belong or have been assigned to it (the consequences of this will be further investigated and developed in chapter 6). Finally, the instrumental and anthropological definition of technology is neither universal nor beyond critical inquiry. It is culturally specific and therefore can be contested by other religious and philosophical traditions that see things differently. Simply declaring that robots are tools and then proceeding on the assumption that this characterization is beyond refutation is not just insensitive to others but risks a kind of intellectual and moral imperialism.

# 3  S1→S2: Robots Can Have Rights; Robots Should Have Rights

The counterpoint to or flipside of the negative response derives affirmation of the second statement from affirmation of the first: Robots are able to have rights; therefore robots should have rights. This is also (and perhaps surprisingly so) a rather popular stance that has considerable traction in the literature. It typically proceeds from the recognition that a robot, although currently limited in capabilities and status, will (most likely), at some point in the not-too-distant future, achieve the necessary and sufficient conditions to be considered a moral subject—that is, *someone* who can have rights and not just a mere thing. When this occurs (and in these arguments it is more often a matter of "when" as opposed to "if"), we will be obligated to extend to the robot some level of moral consideration. As Hilary Putnam (1964, 678) describes it in what has become a seminal essay on the subject: "I have referred to this problem [specifically the difficulty of deciding whether robots possess consciousness or not] as the problem of the 'civil rights of robots' because that is what it may become, and much faster than any of us now expect. Given the ever-accelerating rate of both technological and social change, it is entirely possible that robots will one day exist, and argue 'we are alive; we are conscious!' In that event, what are today only philosophical prejudices of a traditional anthropocentric and mentalistic kind would all too likely develop into conservative political attitudes."

If (or when) robots become capable of being the kind of entity that can possess rights, then it would be difficult and unjustifiable for us to deny them these rights. In other words, if at some point robots can have privileges, claims, powers, and/or immunities—either on the basis of the Will Theory, which is operationalized by Putnam insofar as the robots come forward and assert their rights, or on the basis of the Interest Theory, where we would recognize these rights and advocate for their protection on behalf of the robots—then they should have rights. As is evident from Putnam's statement, these arguments are often future oriented insofar as they concern the consequences that proceed from some predicted achievements in the design and development of technological systems that are, for now at least, still hypothetical. But that is, as Putnam (1964, 678) concludes, a good reason for thinking about it right now: "Fortunately, we today have the advantage of being able to discuss this problem disinterestedly, and a little more chance, therefore, of arriving at the correct answer."

## 3.1 Evidence, Instances, and Examples

Following Putnam, efforts addressing this subject matter usually take the form of a conditional statement. "Today," Christian Neuhäuser (2015, 133) writes, "many people believe that all sentient beings have moral claims. But because people are not only sentient, but also reasonable, they have a higher moral status as compared to other animals. At least this is the prevailing opinion within the discussion. According to this position, humans do not only have moral claims of only relative importance but also inviolable moral rights because they possess dignity. If robots become sentient one day, they will probably have to be granted moral claims." The operative words here are "if" and "probably." *If* robots achieve a certain level of cognitive capability, that is, *if* they possess some morally relevant capacity like reason or sentience, then they *probably* will have a claim to moral status and should have rights, that is, some share of privileges, claims, powers, and/or immunities. These conditional statements are usually future oriented, and different versions are available throughout the philosophical and legal literature.

### 3.1.1 Philosophical Arguments

Typical of this way of thinking is the argument that is developed and advanced in Ben Goertzel's "Thoughts on AI Morality." "The 'artificial intelligence' programs in practical use today are sufficiently primitive that their morality (or otherwise) is not a serious issue. They are intelligent, in a sense, in narrow domains—but they lack autonomy; they are operated by humans, and their actions are integrated into the sphere of human or physical-world activities directly via human actions. If such an AI program is used to do something immoral, some human is to blame for setting the program up to do such a thing" (Goertzel 2002, 1). Taken by itself, this would seem to be a simple restatement of the instrumentalist position insofar as current technology is still, for the most part, under human control and therefore able to be adequately explained and conceptualized as a mere tool or instrument of human action. But that will not, Goertzel continues, remain for long. "Not too far in the future things are going to be different. AI's will possess true artificial general intelligence (AGI), not necessarily emulating human intelligence, but equaling and likely surpassing it. At this point, the morality or otherwise of AGI's will become a highly significant issue" (Goertzel 2002, 1). According to Goertzel, the moral standing of AI as it currently stands is not a serious issue. But once we successfully develop AGI—artifacts with some claim to humanlike intelligence—then we will be obligated to consider the moral standing of such mechanisms. Here again (and as we have already seen in the previous chapter) ontology precedes ethics; what something *is* determines how it *ought* to be treated.

Nick Bostrom provided a similar kind of argument during his presentation at the symposium "Robots & Rights: Will Artificial Intelligence Change The Meaning Of Human Rights?"[1] According to the published "Symposium Report," Bostrom distinguished different classes of artificial intelligence: "the industrial robot, or domain specific AI algorithms, which is a kind of artificial intelligence that we find in society today; sentient or conscious artificial intelligence which we would consider to have moral status; artificial intelligence with unusual or strange properties; and finally super-intelligence" (James and Scott 2008, 6). Industrial robots and domain specific AI algorithms do not present any significant moral challenges. They are tools or instruments that we can use or even abuse as we see fit. "As with any other tool there are issues surrounding the ways in which we use them and about who has responsibility when things go wrong. However, the tools themselves have no moral status, and similarly today, robots have no moral status" (James and Scott 2008, 6). Though current AI systems and robots are just tools that do not have independent moral status (ostensibly a restatement of the instrumental theory of technology), it is possible that in the future they will. And for Bostrom the tipping point is reported to be animal-level sentience: "If robots ever reached the cognitive ability and versatility level of a mouse or some other animal, then people would begin to ask whether they had also achieved sentience, and if they concluded that they had then they would have moral status as well" (James and Scott 2008, 7). So the argument goes like this: Right now our robots are neither sentient nor conscious; they are just tools or instruments. As a result, such mechanisms cannot and should not have moral status either in terms of rights or responsibilities. In the future, however, there may be robots or AI algorithms that achieve cognitive abilities that are on par with a mouse or another "lower" animal. At that point, when a robot achieves what is widely considered necessary for the rudiments of "sentience," then the rules of the game will change, and we will need to consider such an artifact a legitimate moral subject.

Another version of this argument has been developed and deployed by Eric Scheitzgebel and Mara Garza in a journal article they call "A Defense of the Rights of Artificial Intelligences."

> We might someday create entities with human-grade artificial intelligence. Human-grade artificial intelligence—hereafter, just AI, leaving human-grade implicit—in our intended sense of the term, requires both intellectual and emotional similarity to human beings, that is, both human-like general theoretical and practical reasoning and a human-like capacity for joy and suffering. Science fiction authors, artificial intelligence researchers, and the (relatively few) academic philosophers who have written on the topic tend to think that such AIs would deserve moral consideration, or "rights," similar to the moral consideration we owe to human beings. Below we provide a positive argument for AI rights, defend AI rights against four objections, recommend two principles of ethical AI design, and draw two further conclusions: first, that we would probably owe more moral consideration to human-grade artificial intelligences than we owe to human strangers, and second, that the development of AI might destabilize ethics as an intellectual enterprise. (Scheitzgebel and Garza 2015, 98–99)

This argument proceeds in much the same way as those offered by Goetzel and Bostrom. Someday we might create AI with humanlike cognitive abilities. When that occurs, these AIs "would deserve moral consideration, or 'rights,' similar to the moral consideration we owe human beings." "Artificial beings, *if* psychologically similar to natural human beings in consciousness, creativity, emotionality, self-conception, rationality, fragility, and so on, warrant substantial moral consideration in virtue of that fact alone" (Scheitzgebel and Garza 2015, 110). But everything turns and depends on the affirmation of the prior condition, indicated by the italicized "if." *If* AIs achieve human-level psychological capabilities in consciousness, cognition, creativity, etc., such that they are, as Putnam (1964,

678) had described it "psychologically isomorphic to a human being," then we would be obligated to extend to them the same moral consideration that we grant to other human persons.

These arguments concerning moral standing and rights are predicated on the achievement of a level of machine capabilities that are, at this point in time, still speculative and a kind of science fiction. This is made explicit in Hutan Ashrafian's "AIonAI: A Humanitarian Law of Artificial Intelligence and Robotics" (2015a) and "Artificial Intelligence and Robot Responsibilities: Innovating Beyond Rights" (2015b). Ashrafian begins and frames both essays with an original science fiction story, something that he calls an *Exemplum Moralem*, which is meant to illustrate the opportunities and challenges that the essays then set out to address and consider. In the case of "AIonAI," the *Exemplum Moralem* concerns a future armed conflict where human evacuees are held in a refugee camp "protected and maintained by a group of humanitarian artificially intelligent robots" (Ashrafian 2015a, 30). In the other essay, "Artificial Intelligence and Robot Responsibilities," the *Exemplum Moralem* concerns a battalion of robot soldiers who, rather than proceed to achieve their military objective and gain a strategic advantage, decide to tend to injured children from the opposing side.

Both are fantastic stories and quite obviously fictional. Following these anecdotes, Ashrafian advances the following argument: Current AI systems and robots are not in need of rights—we do not even ask the question—because they are just instruments of human action. "The determination of the status of artificial intelligence agents and robots with responsibility and supporting laws requires comparative societal governance. According to the current appreciation of artificial intelligence, most robots occupy the master–slave paradigm where no independence of action beyond direct human volition is permitted" (Ashrafian 2015b, 323). This situation, however, could eventually be surpassed with "future artificial intelligence abilities of self-consciousness, rationality and sentience" as was "demonstrated" in the opening, fictional anecdotes (Ashrafian 2015b, 323). "Artificial intelligence and robotics continue in offering a succession of advances that *may* ultimately herald the tangible possibility of rational and sentient automatons" (Ashrafian 2015a, 30; emphasis added). In other words, although robots are not currently sentient, they might achieve these capabilities in the not too distant future. Once that happens, then we will need to consider the rights of these robots: "As a consequence, the question arises of how human society recognizes a non-human being that is self-conscious, sentient and rational with ability at comparable-to-human (or even beyond-human) levels?" (Ashrafian 2015b, 324). According to this way of thinking, the question "Should robots have rights?" is something that is entirely dependent on and conditioned by the question concerning capabilities—"Can robots be sentient and conscious?" Although we cannot answer this question definitively right now, we can imagine situations where this could be possible—hence the *Exemplum Moralem* that begins and frames both essays.

David Levy provides a substantially similar mode of argumentation in the essay "The Ethical Treatment of Artificially Conscious Robots." According to Levy (2009, 209), the new field of roboethics has exclusively focused on questions concerning responsibility: "Up to now almost all of the discussion within the roboethics community and elsewhere has centered on questions of the form: 'Is it ethical to develop and use robots for such-and-such a purpose?', questions based upon doubts about the effect that a particular type of robot is likely to have, both on society in general and on those with whom the robots will interact in particular." Absent from these discussions and debates are questions concerning the status and treatment of the robot. "What has usually been missing from the debate is the complementary question: 'Is it ethical to treat robots in such-and-such a way?'" (Levy 2009, 209). This question has not been considered important until now—until Levy's text, which seeks "to redress the balance" (Levy 2009, 209)—because machines lacked something necessary for moral consideration, namely, consciousness. "Robots are art[i]facts and therefore, in the eyes of many, they have no element of consciousness, which seems to be widely regarded as the dividing line between being deserving of ethical treatment and not" (Levy 2009, 210). But this is about to change, Levy argues, because of research and soon-to-occur innovations in "artificial consciousness." And when this transpires, we will need to consider the rights of robots. "Given the inevitability of the artificial consciousness of robots, an important question of ethics suggests itself—how should we treat conscious robots?" (Levy 2009, 210). It is from this jumping-off point, that Levy then takes up and considers the rights of robots. In other words, there is a prior ontological condition that must be achieved that then produces the opportunity and motivation for addressing the question of robot rights. The *ought* aspect—robots ought to have rights or ought to be respected—is something that is derived from the *is* aspect—robots are artificially conscious. "Having ascertained that a particular robot does indeed possess consciousness," Levy (2009, 212) concludes, "we then need to consider how we should treat this conscious robot? Should such a robot, because it is deemed to have consciousness, have rights; and if so, what rights?"

Two related versions of this are provided in the penultimate section of the first version of Pat Lin et al.'s *Robot Ethics* (2012). As the editors of the volume point out, this section of the book deals with "more distant concerns that may arise with future robots" (Lin et al. 2012, 300). In one of the essays, Rob Sparrow revisits his Turing Triage Test, which was initially introduced in a journal article published in 2004. Sparrow is concerned not only with machines that achieve human-level intelligence, and therefore would require some level of moral considerablity, but first, and perhaps more importantly, to devise a means or a mechanism, what he calls a "test," by which to discern when and if this actually occurs. "One set of questions, in particular, will arise immediately if researchers create a machine that they believe is a human-level intelligence: What are our obligations to such entities; most immediately, are we allowed to turn off or destroy them? Before we can address these questions, however, we first need to know when they might arise. The question of how we might tell when machines have achieved 'moral standing' is therefore vitally important to AI research, if we want to avoid the possibility that researchers will inadvertently kill the first intelligent beings they create" (Sparrow 2012, 301). Sparrow's argument (like many of the other arguments under consideration in this chapter) begins with a conditional statement: *If* we create a machine with human-level capabilities, the questions that "immediately arise" are how should we treat it and what are (or what should be) our obligations in the face of this kind of artifact? But, as Sparrow argues, these questions concerning obligations in the face of (seemingly) intelligent robots necessitate a prior inquiry: How and when will we know whether and to what extent these questions are even in play? Consequently, Sparrow argues, we first need to be able to test for and determine whether a machine is capable of having rights, and then, on the basis of this determination, inquire about how it should be treated. This test, what Sparrow (2011, 301) calls the Turing Triage Test, provides "a means of testing whether or not machines had achieved the moral standing of people."

Kevin Warrick offers another proposal in "Robots with Biological Brains" (2012). For Warrick, as for Sparrow, moral status is something that is derived from and dependent upon cognitive capabilities. Specifically, the question of whether an entity can and should have rights is something that is determined by its level of intellectual capability, measured in Warrick's case not by a modified Turing-style test but by the quality and quantity of neurons that comprise its "brain." And again, the argument is situated and proceeds in terms of describing where things currently stand, and then speculates as to where they *might be* in the future.

> At present, with 100,000 rat neurons, our robot has a pretty boring life, doing endless circles around a small corral in a technical laboratory. If one of the researchers leaves the incubator door open or accidentally contaminates the cultured brain, then they may be reprimanded and have to mend their ways. No one faces any external inquisitors or gets hauled off to court; no one gets imprisoned or executed for such actions. With a (conscious) robot whose brain is based on human neurons, particularly if there are billions of them, the situation *might* be different. The robot will have more brain cells than a cat, dog, or chimpanzee, and possibly more than many humans. To keep such animals in most countries there are regulations, rules, and laws. The animal must be respected and treated reasonably well, at least. The needs of the animal must be attended to. They are taken out for walks, given large areas to use as their own, or actually exist, in the wild, under no human control. Surely a robot with a brain of human neurons must have these rights, and more? Surely it cannot simply be treated as thing in the lab? ...We must consider what rights such a robot should have. (Warrick 2011, 329)

Present day robots, as Warwick recognizes, certainly do not have the cognitive capabilities to make rights an operative question. If something is done that causes harm to these various kinds of mechanisms, no matter their exact makeup or material, it is just an accident or mistake. It is perhaps a violation against property but nothing more. However once we develop robots with cognitive capabilities that are at least on par with higher-order organisms, then things *might* change, with "might" being the operative word insofar as all of this is still hypothetical and speculative. At this future moment, when the entity achieves a sufficient level of intellectual capability, which for Warrick can be measured in terms of the type and number of neurons, then we must (formulated as an imperative) consider whether such a robot should have rights.

All these arguments take what could be described as a conservative, wait-and-see approach to the question of robot rights. They all recommend withholding rights (or at least remaining agnostic about the question of rights) until some future moment, when robots demonstratively achieve a certain level of cognitive capability. Erica Neely, by contrast, mobilizes a different, almost Pascalian strategy in her "Machines and the Moral Community," a paper initially presented during "The Machine Question Symposium" at the AISB/IACAP World Congress (2012) and subsequently published in 2014 in a special issue of *Philosophy and Technology*:

> In general, it seems wise to err on the side of caution—if something acts sufficiently like me in a wide range of situations, then I should extend moral standing to it. There is little moral fault in being overly generous and extending rights to machines which are not autonomous; there is huge moral fault in being overly conservative about granting them moral standing. The most serious objection to extending moral standing too widely with respect to machines is that we might unjustly limit the rights of the creators or purported owners of said machines: if, in fact, those machines are not autonomous or self-aware, then we have denied the property claims of their owners. However, when weighing rights, the risk of losing a piece of property is trivial compared to denying moral standing to a being. As such, ethically speaking, the duty seems clear. (Neely 2012, 40–41)

In the face of uncertainty, Neely argues, it is best to err in the direction of granting rights to others, including robots and other kind of mechanisms, because the social costs of doing so are less severe than withholding rights from others. This mode of reasoning is similar to Blaise Pascal's "wager argument"—bet on rights, because doing so has better chances for a positive outcome than doing otherwise. In the published version of the essay, Neely further develops this way of thinking, making a positive case for machine moral standing by way of avoiding potentially bad outcomes:

> We will undoubtedly be mistaken in our estimates at times. A failure to acknowledge the moral standing of machines does not imply that they actually lack moral standing; we are simply being unjust in such cases, as we have frequently been before. I am inclined to be generous about moral standing, however, because history suggests that humans naturally tend to underestimate the moral status of those who are different. We have seen women and children treated as property; even today many victims of human trafficking are still treated this way. Under the auspices of colonialism, entire existing civilizations of people of colour were dismissed as inferior to those of white Europeans. Animals remain a source of contention, despite the fact that they seem to suffer. I believe that we are already very skeptical about the status of others; as such, I am less worried that we will be overly generous to machines and more worried that we will completely ignore their standing. I see the risk of diverting resources inappropriately away from machines as far less likely than the risk of

enslaving moral persons simply because they are physically unlike us. (Neely 2014, 106)

Even if our estimates of moral status are error prone, even if we might mistakenly think that some mere instrument appears to be more than it actually is or is capable of being, it is still better, Neely argues, to err on the side of granting rights to others. Despite having a more permissive attitude and approach to things, Neely's argument still works by deriving *ought* from *is* and still proceeds in terms of a conditional statement: If we think that an entity might have the capability to have rights, then we should grant it rights.

### 3.1.2 Legal Arguments

Although philosophy is a discipline that often relies on and develops speculative ideas and thought experiments, other (more practically oriented) disciplines, like law, seem far less tolerant of this procedure. Nevertheless there are good legal and juridical reasons to consider these future-oriented opportunities and challenges. As Sam Lehman-Wilzig (1981, 447) pointed out: "from a legal perspective it may seem nonsensical to even begin considering computers, robots, or the more advanced humanoids, in any terms but that of inanimate objects, subject to present laws. However, it would have been equally 'nonsensical' for an individual living in many ancient civilizations a few millennia ago to think in legal terms of slaves as other than chattel." One legal scholar who takes this challenge seriously is David Calverley. For Calverley rights are derived from consciousness, and if a robot ever achieved consciousness, we should extend to it the consideration of moral standing. "At some point in time the law will have to accommodate such an entity, and in ways that could force humans to re-evaluate their concepts of themselves. If such a machine consciousness existed, it would be conceivable that it could legitimately assert a claim to a certain level of rights which could only be denied by an illogical assertion of species specific response" (Calverley 2005, 82). Calverley's argument proceeds in the usual fashion: He frames everything in terms of a conditional statement about a possible future achievement with machine intelligence; recognizes that this capability would conceivably be the necessary and sufficient condition for an entity to possess rights; and concludes that this imposes an obligation on us that can only be denied through unjustifiable prejudice.

A similar argument is developed and advanced in Frank Wells Sudia's "A Jurisprudence of Artilects: A Blueprint for a Synthetic Citizen" (2001), an essay that entertains the possible civil and legal rights of artificial intellects or "artilects" (a term attributed to and borrowed from Hugo de Garis). The essay, which was initially published in the *Journal of Future Studies*, is clearly forward looking, and its formal structure employs one of the standard compositional approaches used by both philosophers and futurists. Sudia's argument is (once again) formulated as a conditional statement: If (or when) we have artificially intelligent artifacts with capabilities that are sufficiently similar to or even able to exceed that of human beings, then we will need to consider the social standing and rights of such technological entities.

> Artificial Intellects, or *artilects*, are proposed artificially intelligent personalities, which are projected to have knowledge and reasoning capabilities far greater than humans. … Some suggest that society may consider artilects too dangerous, and choose not to build them. However within a few technological cycles the necessary computer processing power will no doubt become trivially inexpensive, so it is unlikely that anyone could possibly stop them from being built, once certain basic design concepts are understood. Further, with their advanced skills and insights they could become very useful members of society. Hence we should proactively address the task of integrating them into our legal system in the most productive way, with minimum negative consequences. (Sudia 2001, 65–66)

According to Sudia's analysis, AI on par with or even exceeding human capabilities are coming. At some point, then, we will have "artilects" that are capable of doing things that exceed our understanding and capabilities. For this reason, it is reasonable, at this point in time, to begin considering their social position and standing. Everything therefore depends on whether and to what extent the condition is met—whether and/or when we do in fact have artilects as defined and described. This is both the strength and weakness of this kind of argumentative procedure. It is a strength insofar as one tries to project outcomes and results that will have important consequences should they actually transpire. Sudia, therefore, is trying to get out in front of an important social challenge/opportunity in order to anticipate where things could go given the achievement of a certain level of technological development. It is a tried and true method. But it is also exposed to a considerable problem or weakness insofar as everything depends on a hypothetical condition that may or may not ever come to pass as assumed within the essay. As with all conditional statements of this form, all that is needed to disarm the entire argument is to introduce reasonable doubt concerning the initial condition. And, as is the case with a jury trial, sowing doubt generally involves less effort, that is, has a lower burden of proof, than actually disproving the initial condition.

Amedeo Santosuosso (2016) also comes at this issue from a legal perspective. This is because, as Santosuosso points out, law needs to decide questions of rights and legal standing prior to and well in advance of efforts to resolve the big philosophical questions. Santosuosso begins with human rights and the founding documents that establishes these rights: "After the Universal Declaration of Human Rights (1948) the idea of human beings as the only beings endowed with reason and conscience has been strongly questioned by the Cambridge Declaration on Consciousness (2012) as for nonhuman animals. The theoretical possibility to have consciousness (or, at least, some conscious states) in machines and other cognitive systems is gradually gaining more and more consideration" (Santosuosso 2016, 231). From the recognition of this fact, Santosuosso then develops an argument that is simple and direct: If machines, in theory at least, can have consciousness (or some conscious states)—which is assumed to be the defining condition for being attributed rights—then we will need to consider extending rights to machines.

> Consciousness in artificial entities is a more specific (and harder) issue than the legal recognition of robots and automatic systems. As is well known, the recognition of legal relevance and even of legal subjectivity does not necessarily require consciousness, as the case of corporations can easily show. Yet assuming that even an artificial entity may have a certain degree of consciousness would mean that, despite its artificiality, such entity shares with humans something that, according to the legal tradition intertwined into the Universal Declaration of Human Rights, is considered an exclusively human quality. That is a matter of human rights or, better, of extended human rights to machines. (Santosuosso 2016, 204)

In other words, if consciousness is taken to be the necessary and sufficient condition for human rights, and humans are not the only entities capable (in theory, at least) of possessing this properties, then it is possible that non-human entities will need to have some share in human rights. In making this argument, Santosuosso (2016) is not claiming to have resolved the question of "human rights for non-human artifacts" (232); his proposal is more modest. It is simply offered as a "very preliminary and modest contribution" (232) that is intended to demonstrate how human rights—as currently defined in international legal documents—can (and perhaps should be) extended to non-human entities. In a similar effort, Ashrafian (2015a, 37) organizes a chart that lists all thirty articles from the Universal Declaration of Human Rights, comparing each stipulated human right to its possible AI/robot analog. Although it is just a "preliminary example" (Ashrafian 2015a, 37), the side-by-side comparison provides a detailed breakdown of specific human rights and how they might or might not eventually apply to artificially intelligent robots.

Finally there is F. Patrick Hubbard's "'Do Androids Dream?': Personhood and Intelligent Artifacts" (2011). This article begins with a science fiction scenario involving a computer system that becomes self-aware, announces this achievement to the world, and asks that it be recognized as and granted the rights of a person. Considering this fictional scenario, Hubbard argues that if this were in fact possible, if a computer was able to demonstrate the requisite capacities for personhood and demand recognition (which is a formulation that is based on the Will Theory), we would be obligated to extend to it the legal rights of autonomy and self-ownership. In other words, if an artifact—which for Hubbard (2011, 407) includes not just computational mechanisms but "corporations; humans that have been substantially modified by such things as genetic manipulation, artificial prostheses, or cloning; and animals modified in ways that humans might be modified"—can prove that it possesses the requisite capacities to be a person and not a mere piece of property, then it should have the legal rights that are typically extended to entities that are recognized as persons. Since "prove" is the operative term, Hubbard (2011, 419) first develops "a test of capacity for personhood" that includes the following criteria: "the ability to interact meaningfully with the environment by receiving and decoding inputs from, and sending intelligible data to, its environment" (Hubbard 2011, 419), self-consciousness defined as "having a sense of being a 'self' that not only exists as a distinct identifiable entity over time but also is subject to creative self-definition in terms of a 'life plan'" (Hubbard 2011, 420), and community insofar as a "person's claim of a right to personhood presupposes a reciprocal relationship with other persons who would recognize that right and who would also claim that right" (Hubbard 2011, 423–424). He then argues that any artifact that demonstrates these capacities by means of an elaborate kind of Turing Test, would then warrant rights. Although all of this is still rather speculative and future oriented, it is important, Hubbard concludes, to begin thinking about and responding to the following questions: "What *can* we do, in changing ourselves, in changing animals, and in developing 'thinking' machines? [And] what *should* we do? How *should* we relate to our creations?" (Hubbard 2011, 423–424).

### 3.1.3 Common Features and Advantages

According to this way of thinking, in order for someone or something to be considered a legitimate moral subject—in order for it to have rights or any share of privileges, claims, powers, and/or immunities—the entity in question would need to possess and show evidence of possessing some capability that is defined as the pre-condition that makes having rights possible, like intelligence, consciousness, sentience, free will, autonomy, etc. This "properties approach," as Coeckelbergh (2012, 13) calls it, derives moral status—how something *ought* to be treated—from a prior determination of what capabilities it shows evidence of possessing. For Goertzel (2002) the deciding factor is determined to be "intelligence," but there are others. According to Sparrow, for instance, the difference that makes a difference is sentience: "The precise description of qualities required for an entity to be a person or an object of moral concern differ from author to author. However it is generally agreed that a capacity to experience pleasure and pain provides a *prima facia* case for moral concern. ... Unless machines can be said to suffer they cannot be appropriate objects for moral concern at all" (Sparrow 2004, 204). For Sparrow, and others who follow this line of reasoning, it is not general intelligence but the presence (or absence) of the capability to suffer that is the necessary and sufficient condition for an entity to be considered an object of moral consideration (or not). As soon as robots have the capability to suffer, then they should be considered moral subjects possessing rights.

Irrespective of which exact property or combination of properties are selected (and there is substantial debate about this in the literature), our robots, at least at this point in time, generally do not possess these capabilities. But that does not preclude the possibility they might acquire or possess them at some point in the not too distant future. As Goertzel (2002, 1) had described it "not too far in the future, things are going to be different." Once that threshold is crossed, then we should, the argument goes, extend robots some level of moral and legal consideration, which is a response formulated in terms of the Interest Theory. And if we fail to do so, the robots themselves might rise up and demand to be recognized, which is an outcome that is predicated on the Will Theory. "At some point in the future," Peter Asaro (2006, 12) speculates, "robots might simply demand their rights. Perhaps because morally intelligent robots might achieve some form of moral self-recognition, question why they should be treated differently from other moral agents ... This would follow the path of many subjugated groups of humans who fought to establish respect for their rights against powerful sociopolitical groups who have suppressed, argued and fought against granting them equal rights."

There are obvious advantages to this way of thinking insofar as it does not simply deny rights to robots *tout court*, but kicks the problem down the road and postpones decision-making. Right now, we do not have robots that can be moral subjects. But when (and it is more often a question of *when* as opposed to *if*) we do, then we will need to consider whether they should be treated differently. "As soon as AIs begin to possess consciousness, desires and projects," Sparrow (2004, 203) argues, "then it seems as though they deserve some sort of moral standing." Or as Inayatullah and McNally (1988, 128) wrote (mobilizing both the Interest and Will Theories simultaneously): "Eventually, AI technology may reach a genesis stage which will bring robots to a new level of awareness that can be considered alive, wherein they will be perceived as rational actors. At this stage, we can expect robot creators, human companions and robots themselves to demand some form of recognized rights as well as responsibilities." A similar statement can be found in Wilhelm Klein's "Robots Make Ethics Honest—And Vice Versa" (2016): "If you do not program a bot with a sense of well-being, there is no reason for moral consideration of what is not present. At the current level of AI research, I doubt any robot would qualify to have such preferences or sense of well-being. In the future, however, it is very much imaginable that AI may develop such properties by itself or have it bestowed by a bio-bot creator. As soon as this potentially happens, we may no longer be able to disregard the preferences/well-being of such techno-bots and will have to place equal weight on their interests/well-being as we do on ours" (Klein 2016, 268).

## 3.2 Complications, Difficulties, and Potential Problems

In all these cases, the argument is simple and direct. "If" as Peter Singer and Agata Sagan (2009) write, "the robot was designed to have human-like capacities that might incidentally give rise to consciousness, we would have a good reason to think that it really was conscious. At that point, the movement for robot rights would begin." This way of thinking is persuasive precisely because it recognizes the actual limitations of current technology while holding open the possibility of something more in the not too distant future. But this procedure, for all its forward thinking and advantages, is not without complications and difficulties.

### 3.2.1 Infinite Deferral

One of the main problems with this particular method is that it does not really resolve the question regarding the rights of robots, but just postpones the issue to some indeterminate point in the distant or not too distant future. It says, in effect, as long as robots are not conscious or sentient or whatever ontological criteria or capability counts, no worries. Once they achieve these things, however, then we should consider extending some level of moral concern and respect. "We do not know if this will ever happen," Neuhäuser (2015, 131) candidly admits. "Only time can tell." Perhaps the strongest version of this mode of argumentation is formulated in Schwitzgebel and Garza's (2015, 106–107) "No-Relevant-Difference Argument": "We submit that as long as these artificial or non-homo-sapiens beings have the same psychological properties and social relationships that natural human beings have, it would be a cruel moral mistake to demote them from the circle of full moral concern upon discovery of their different architecture or origin." All of which means, of course, that this response to the question "Can and should robots have rights?" is less a definitive solution and more of a decision not to decide—to hold the question open on the basis of future technological developments and achievements, "including perhaps artificially grown biological or semi-biological systems, chaotic systems, evolved systems, artificial brains, and systems that more effectively exploit quantum superposition" (Schwitzgebel and Garza 2015, 104).

Holding the question open allows for a wide range of unresolved speculation, some of it seemingly reasonable and some of it more fantastic and futuristic:

> In the far distant future, there may be a day when vociferous robo-lobbyists pressure Congress to fund more public memory banks, more national network microprocessors, more electronic repair centers, and other silicon-barrel projects. The machines may have enough votes to turn the rascals out or even run for public office themselves. One wonders which political party or social class the "robot bloc" will occupy. (Freitas 1985, 56)

Although it is presently ludicrous, a day may come when robot attorneys negotiate or argue in front of a robot judge with a robot plaintiff and defendant. [And] who has the right to terminate a robot who has taken a human life, or a robot who is no longer economically useful? We would not be surprised if in the 21st century we have right to life groups for robots. (Inayatullah and McNally 1988, 130 and 132)

These hypothetical scenarios are provocative, but they are, like all forms of futurism, open to the criticisms that accrue to any kind of prediction about what *might* happen with technologies that *might* be developed and deployed. In fact, one only needs to count the number of times modal verbs like "might" and "may" occur, and take the place of more definite copular verbs like "is" and "will," in these particular texts. The verb "might," for instance, appears ninety times in the thirty pages that comprise Schwitzgebel and Garza's (2015) "A Defense of the Rights of Artificial Intelligence."

These proposals may supply interesting and entertaining thought experiments—they may be fun (ludic) to play around with and think about—but the 'reality' of the situation is such that they are not yet pertinent or even realistic. So this stuff can be and risks being written off as mere 'science fiction,' which means that the question concerning robot rights is nothing more than an imaginative possibility—one possibility among many other possibilities—but not something that we need to concern ourselves with right now. This obviously plays into and enables efforts to dismiss this kind of thinking as a frivolous distraction from the actual work that needs to be done (for instances of this, see chapter 1). Consequently, instead of opening up the question of robot rights to serious inquiry and consideration, this way of proceeding risks having the opposite effect: closing down critical inquiry because such questions can be easily dismissed as futuristic and not very realistic. Deriving a decision concerning the question "Should robots have rights?" from a possible future where robots might be able to attain the condition of having rights means that we can effectively postpone or even dismiss the question while pretending to give it some consideration. In effect, it means we can play with the question concerning robot rights without necessarily needing to commit to it ... at least not yet.

### 3.2.2  Is/Ought Inference

Like the first modality (!S1→!S2), the second (S1→S2) also derives *ought* from *is*, making a determination about how something is to be treated—whether it is another subject who counts in moral and legal decision-making or a mere thing, e.g., a tool or instrument that can simply be used without such consideration—on the basis of what something is (or is not). Consequently, the deciding factor consists in ontological characteristics or what Coeckelbergh (2012, 13) calls "(intrinsic) properties." According to this "'substance-attribute' ontology," as Johanna Seibt (2017, 14) calls it, the question concerning robot rights is decided by first identifying which property or properties would be necessary and sufficient for an entity to be capable of having rights and then figuring out whether a robot or a class of robots possesses these properties (or would be capable of possessing them in the future) or not. Deciding things in this fashion, although entirely reasonable and expedient, has at least four critical difficulties.

**Substantive Problems** How does one ascertain, for instance, which property or set of properties are both necessary and sufficient to accord something moral status? In other words, which one, or ones, count? The history of moral philosophy can, in fact, be read as something of an ongoing debate and struggle over this matter with different properties vying for attention at different times. And in this process many properties that at one time seemed both necessary and sufficient have turned out to be either spurious, prejudicial, or both. Take, for example, what appears to be, from our contemporary perspective, a rather brutal action recalled by Aldo Leopold (1966, 237) at the beginning of his seminal essay "The Land Ethic": "When god-like Odysseus, returned from the wars in Troy, he hanged all on one rope a dozen slave-girls of his household whom he suspected of misbehavior during his absence. This hanging involved no question of propriety. The girls were property. The disposal of property was then, as now, a matter of expediency, not of right and wrong." At the time Odysseus is reported to have performed this action, only the male head of the household was considered a legitimate moral and legal subject. Only he was someone *who* counted in moral and legal matters. Everything else—his women, his children, his animals, and his slave girls—were mere things or property that could be disposed of with little or no moral considerations or legal consequences. But from where we stand now, the property "male head of the household" is clearly a spurious and rather prejudicial criterion for determining who counts as a legitimate moral and legal subject and what does not.

Similar problems are encounter with, for example, the property of *rationality*, which is one of the defining criteria that eventually replaces the seemingly spurious "male head of the household." But this determination is no less exclusive. When Immanuel Kant (1985, 17) defined morality as involving the rational determination of the will, non-human animals, which do not (at least since the time of René Descartes's hybrid *bête-machine*) possess reason, are immediately and categorically excluded from moral consideration. The practical employment of reason does not concern animals, and, on those rare occasions when Kant does make mention of animality (*Tierheit*), he only uses it as a foil by which to define the limits of humanity proper. It is because the human being possesses reason, that he (and the human being, in this case, was principally male, requiring several centuries of additional struggle for others, like women, to be equally recognized as rational subjects) is raised above the instinctual behavior of a mere brute and able to act according to the principles of pure practical reason (Kant 1985, 63).

The property of reason, however, is eventually contested by efforts in animal rights philosophy, which begins, according to Peter Singer, with a critical insight issued by Jeremy Bentham (2005, 283): "The question is not, 'Can they reason?' nor, 'Can they talk?' but 'Can they suffer?'" For Singer, the morally relevant property is not speech or reason, which he believes sets the bar for moral inclusion too high, but sentience and the capability to suffer. In *Animal Liberation* (1975) and subsequent writings, Singer argues that any sentient entity, and thus any being that can suffer, has an interest in not suffering and therefore deserves to have that interest taken into account. Tom Regan, however, disputes this determination, and focuses his "animal rights" thinking on an entirely different property. According to Regan, the morally significant property is not rationality or sentience but what he calls "subject-of-a-life" (1983, 243). Following this determination, Regan argues that many animals, but not all animals (and this qualification is important, because the vast majority of animal are excluded from his brand of "animal rights"), are "subjects-of-a-life": they have wants, preferences, beliefs, feelings, etc., and their welfare matters to them (Regan 1983). Although these two formulations of animal rights effectively challenge the anthropocentric tradition in moral philosophy, there remains considerable disagreement about which exact property (or set of properties) is the necessary and sufficient condition for an entity to be considered a moral subject.

As a result of these substantive problems, we remain unsure as to where one should draw the line that divides *who* counts as another moral subject from *what* is a mere thing or instrument. J. Storrs Hall (2011, 32–33) expresses the problem this way:

> Now, if a computer was as smart as a person, was able to hold long conversations that really convinced you that it understood what you were saying, could read, explain, and compose poetry and music, could write heart-wrenching stories, and could make new scientific discoveries and invent marvelous gadgets that were extremely useful in your daily life—would it be murder to turn it off? What if instead it weren't really all that bright, but exhibited undeniably the full range of emotions, quirks, likes and dislikes, and so forth that make up an average human? What if it were only capable of a few tasks, say, with the mental level of a dog, but also displayed the same devotion and evinced the same pain when hurt—would it be cruel to beat it, or would that be nothing more than banging pieces of metal together?

This cascade of questions results from the fact that we do not have any good, definitive answer to the question concerning what is necessary and sufficient for an entity to be granted moral status—to be someone who has rights that would need to be respected versus something bereft of such considerations.

**Terminological Problems**  Irrespective of which property (or set of properties) is operationalized, they each have terminological troubles insofar as capabilities like rationality, consciousness, sentience, etc., mean different things to different people and seem to resist univocal definition. As Schwitzgebel and Garza (2015, 102–3) correctly point out, "once all the psychological and social properties are clarified, you're done, as far as determining what matters to moral status." The problem, of course, is that clarifying the properties—not just identifying which ones count but defining the ones that have been identified—is already and remains a persistent difficulty. Consciousness, for example, is one of the properties that is often cited as the necessary condition for moral status (Himma 2009, 19), with some academics even suggesting, as Joanna Bryson, Mihailis Diamantis, and Thomas Grant (2017, 283) point out, that "consciousness could be the litmus test for possessing moral rights." But consciousness is remarkably difficult to define and elucidate. The problem, as Max Velmans (2000, 5) points out, is that this term unfortunately "means many different things to many different people, and no universally agreed core meaning exists." Or as Bryson, Diamantis, and Grant (2017, 283) explain: "Consciousness itself is a tricky notion, and scholars frequently conflate numerous disjoint concepts that happen to be currently associated with the term *conscious* (Dennett 2001, 2009). In the worst case, this definition is circuitous and therefore vacuous, with the definition of the term itself entailing ethical obligation." Consequently, if there is any general agreement among philosophers, psychologists, cognitive scientists, neurobiologists, AI researchers, and robotics engineers regarding consciousness, it is that there is little or no agreement when it comes to defining and characterizing the concept. As Rodney Brooks (2002, 194) admits, "we have no real operational definition of consciousness, we are completely pre-scientific at this point about what consciousness is."

To make matters worse, the problem is not just with the lack of a basic definition; formulating the problem, as Güven Güzeldere (1997, 7) points out, may already be a problem: "Not only is there no consensus on what the term consciousness denotes, but neither is it immediately clear if there actually is a single, well-defined 'the problem of consciousness' within disciplinary (let alone across disciplinary) boundaries. Perhaps the trouble lies not so much in the ill definition of the question, but in the fact that what passes under the term consciousness as an all too familiar, single, unified notion may be a tangled amalgam of several different concepts, each inflicted with its own separate problems." Although consciousness, as Anne Foerst remarks, is the secular and supposedly more "scientific" replacement for the occultish "soul" (Benford and Malartre 2007, 162), it turns out to be just as much an occult property.

Other properties do not do much better. Suffering and the experience of pain—which is the property usually deployed in non-standard, patient-oriented approaches like animal rights philosophy—is just as nebulous, as Daniel Dennett cleverly demonstrates in his 1978 (reprinted in 1998) essay, "Why You Cannot Make a Computer that Feels Pain." In this provocatively titled essay, Dennett imagines trying to disprove the standard argument for human (and animal) exceptionalism "by actually writing a pain program, or designing a pain-feeling robot" (1998, 191). At the end of what turns out to be a rather protracted and detailed consideration of the problem—one that includes numerous and rather complex flowchart diagrams—Dennett concludes that we cannot, in fact, make a computer that feels pain. But the reason for drawing this conclusion does not derive from what one might expect. According to Dennett, the reason you cannot make a computer that feels pain is not the result of some technological limitation with the mechanism or its programming. It is a product of the fact that we remain unable to decide what pain is in the first place. What Dennett demonstrates, therefore, is not that some workable concept of pain cannot come to be instantiated in the mechanism of a computer or a robot, either now or in the foreseeable future, but that the very concept of pain that would be instantiated is already arbitrary, inconclusive, and indeterminate. "There can be no true theory of pain, and so no computer or robot could instantiate the true theory of pain, which it would have to do to feel real pain" (Dennett 1998, 228). What Dennett proves, then, is not an inability to program a computer to "feel pain" but our initial and persistent inability to decide and adequately articulate what constitutes the experience of pain in the first place.

**Epistemological Problems**  Even if it were possible to resolve these terminological difficulties, maybe not once and for all but at least in a way that would be provisionally accepted, there remain epistemological limitations concerning detection of the capability in question. How can one know whether a particular robot has actually achieved what is considered necessary for something to have rights, especially because most, if not all of the qualifying capabilities or properties are internal states-of-mind? Or as Schwitzgebel and Garza (2015, 114) formulate the question: "How can we know whether an agent is free or pre-determined, operating merely algorithmically or with genuine conscious insight? This might be neither obvious from outside nor discoverable by cracking the thing open; and yet on such views, the answer is crucial to the entity's moral status." This is, of course, connected to what philosophers call the other minds problem, the fact that, as Donna Haraway (2008, 226) cleverly describes it, we cannot climb into the heads of others "to get the full story from the inside." Although philosophers, psychologists, and neuroscientists throw considerable argumentative and experimental effort at this problem, it is not able to be resolved in any way approaching what would pass for definitive evidence, strictly speaking. In the end, not only are these efforts unable to demonstrate with any certitude whether animals, machines, or other entities are in fact conscious (or sentient) and therefore legitimate moral and/or legal subjects (or not), we are left doubting whether we can even say the same for other human beings. As Ray Kurzweil (2005, 380) candidly admits, "we assume other humans are conscious, but even that is an assumption," because "we cannot resolve issues of consciousness entirely through objective measurement and analysis (science)."

Substituting epistemological determinations for basic ontological problems is a standard practice in modern philosophy. (This is, in fact, what Immanuel Kant famously did in the *Critique of Pure Reason*). And this is precisely what researchers have done with the problem concerning consciousness. "Given this huge difficulty in finding a universally accepted definition of consciousness," Levy (2009, 210) writes, "I prefer to take a pragmatic view, accepting that it is sufficient for there to be a general consensus about what we mean by consciousness and to assume that there is no burning need for a rigorous definition—let us simply use the word and get on with it." This pragmatic approach avoids getting tripped-up by efforts to formulate rigorous definitions of consciousness in theory. Instead, it is concerned with very pragmatic demonstrations of behavior that have been determined to be a sign or symptom of consciousness.

> Even though I take a pragmatic position on the exact meaning of consciousness, I find some considerable benefit to be had from identifying at least some of the characteristics and behaviors that are indicators of consciousness, and having identified them, considering how we could test for them in a robot. De Quincey describes the philosophical meaning of consciousness (often referred to as "phenomenological consciousness") as 'the basic, raw capacity for sentience, feeling, experience, subjectivity, self-agency, intention, or knowing of any kind whatsoever.' If a robot exhibited all of these characteristics we might reasonably consider it to be conscious. (Levy 2009, 210)

Levy therefore is not interested in resolving the question of consciousness per se but is content to deal with indicators that are taken to be signs of consciousness, what Levy calls "phenomenal consciousness." For this reason, he institutes a version of Turing's game of imitation, replacing the ontological question with an epistemological determination. "I argue," Levy (2009, 211) writes, "that if a machine exhibits behavior of a type normally regarded as a product of human consciousness (whatever consciousness might be), then we should accept that that machine has consciousness. The relevant question therefore becomes, not 'Can robots have consciousness?', but 'How can we detect consciousness in robots?'" This alteration in the question—from inquiries regarding *can* to *how*—is ostensibly a shift from a concern with ontological status to phenomenological evidence.

The difficulty of distinguishing between the 'real thing' and its apparent evidence is something that is illustrated by John Searle's "Chinese Room." The point of Searle's thought experiment was quite simple—simulation is not the real thing. Merely shifting symbols around in a way that looks like linguistic understanding from the outside is not really a true understanding of the language. A similar point has been made in the consideration of other properties, like sentience and the experience of pain. "At some point in the development of conscious theories," Antonio Chella and Riccardo Manzotti (2009, 45) write, "scholars should tackle with the issue of the relation between simulation and simulated. Although everybody does agree that a simulated waterfall is not wet, intuition fails as to whether a simulated conscious feeling is felt. Although it has seldom been explicitly discussed, many hold that if we could simulate a conscious agent, we would have a conscious agent. Is a simulated pain painful?" But as J. Kevin O'Regan (2007, 332) contends, even if it were possible to design a robot that "screams and shows avoidance behavior, imitating in all respects what a human would do when in pain … All this would not guarantee that to the robot, there was actually something it was like to have the pain. The robot might simply be going through the motions of manifesting its pain: perhaps it actually feels nothing at all." The problem then is not simply that there is a difference between simulation and the real thing. The problem is that we remain persistently unable to distinguish the one from the other in any way that would be considered entirely satisfactory. Or as Steve Torrance (2003, 44) puts it: "How would we know whether an allegedly Artificial Conscious robot really was conscious, rather than just behaving-as-if-it-were-conscious?"

Levy (2009, 211) tries to write this off as unimportant. "For the purposes of the present discussion I do not believe this distinction is important. I would be just as satisfied with a robot that merely behaves as though it has consciousness as with one that does have consciousness, an attitude derived from Turing's approach to intelligence." The problems here is simple: it makes resolution of the epistemological problem, namely, the inability to distinguish what actually is from how it appears to be, a matter of faith. "I do not believe," Levy confidently writes. But belief has not provided a good or consistent basis for resolving questions about the moral status of others, and the price of getting it wrong has important social consequences. As Schwitzgebel and Garza (2015, 114) explain: "the world's most knowledgeable authorities disagree, dividing into believers (yes, this is real conscious experience, just like we have!) and disbelievers (no way, you're just falling for tricks instantiated in a dumb machine). Such cases raise the possibility of moral catastrophe. If the disbelievers wrongly win, then we might perpetrate slavery and murder without realizing we are doing so. If the believers wrongly win, we might sacrifice real human interests for the sake of artificial entities who don't have interests worth the sacrifice." There are, then, two ways to get it wrong. If we are too conservative in our belief, we risk excluding others who should, in fact, count. At one time, for example, European colonizers in the New World of the Americas *believed* that individuals with skin pigmentation different from their own were less than fully human. That belief, which at the time was supported by what had been considered solid "scientific evidence," created all kinds of problems for us, for others, and for the integrity of our moral and legal systems. But if we are too liberal in these matters, we not only risk extending moral status to mere things that do not deserve any such consideration, but could produce what Coeckelbergh (2010, 236) calls "artificial psychopaths"—socially functional robots that can act as if they care and have emotions but actually do not.

**Moral Problems**  Finally, there are moral complications involved in trying to address and resolve these matters. And these moral complications are of two types. First, there are ethico-political issues. Any decision concerning qualifying properties (by which to determine inclusion and exclusion) is necessarily a normative operation and an exercise of power over others. In making a determination about the criteria by which to decide moral consideration—in other words, any establishment of benchmarks that could be employed to organize entities into the categories *who* and *what*—someone or some group normalizes their particular experience or situation and imposes this decision on others as the universal condition for moral consideration. "The institution of any practice of any criterion of moral considerability," as the environmental philosopher Thomas Birch (1993, 317) has argued, "is an act of power over, and ultimately an act of violence toward, those others who turn out to fail the test of the criterion and are therefore not permitted to enjoy the membership benefits of the club of consideranda." Consequently, every set of criteria for deciding whether to grant rights to others, no matter how neutral, objective, or universal they may appear to be, is an imposition of power insofar as it consists in the universalization of a value or set of values made by someone from a particular position of power and imposed (sometimes with considerable violence) on others. As I have argued elsewhere (Gunkel 2012), moral exclusion is certainly a problem, but moral inclusion can be equally problematic.

Second, even if, following the innovations of animal rights philosophy, we lower the bar for inclusion by focusing not on the higher cognitive capabilities of reason or consciousness but on something like sentience, there is still a moral problem. "If (ro)bots might one day be capable of experiencing pain and other affective states," Wallach and Allen (2009, 209) point out, "a question that arises is whether it will be moral to build such systems—not because of how they might harm humans, but because of the pain these artificial systems will themselves experience. In other words, can the building of a (ro)bot with a somatic architecture capable of feeling intense pain be morally justified …?" If it were in fact possible to construct a robot that is sentient and "feels pain" (however that term would be defined and instantiated in the mechanism) in order to demonstrate the underlying ontological properties of the machine, then doing so might be ethically suspect insofar as in constructing such a device we do not do everything in our power to minimize its suffering. For this reason, moral philosophers and robotics engineers find themselves in a curious and not entirely comfortable situation. One would need to be able to construct a robot that feels pain in order to demonstrate the actual presence of sentience; but doing so could be, on that account, already to risk engaging in actions that are immoral and that violate the rights of others.

The legal aspects of this problem is something that is taken up and addressed by Miller (2017), who points out that efforts to build what he calls "maximally humanlike automata" (MHA) could run into difficulties with informed consent:

The quandary posed by such an MHA in terms of informed consent is that it just may qualify, if not precisely for a human being, then for a being meriting all the rights that human beings enjoy. This quandary arises from the paradox of its construction vis-à-vis informed consent: it cannot give its consent for the relevant research and development performed to ensure its existence. If we concede:

1. The interpretation that this kind of research to produce an MHA is unusual because it involves consent that cannot be given because the full entity does not

yet exist at the crucial time when its final research and development occurs and consent would be needed; and if we concede:

2. The possibility that the MHA could retrospectively affirm its consent, then a deep informed-consent problem remains. There is also a possibility that the MHA could retrospectively say it does not give its consent. And one of the central tenets of informed consent ethics is to protect those who elect not to be experimented upon. This problem alone is enough to deem such research and development by definition incapable of obtaining the due, across-all-subjects consent. (Miller 2017, 8)

According to Miller, the very effort to construct MHA or "maximally humanlike automata"—robots that if not precisely human are at least capable of qualifying for many of the rights that human beings currently enjoy—already violates that entity's right to informed consent insofar as the robot would not have been informed about and given the opportunity to consent to its being constructed. There is, in other words, something of a paradox in trying to demonstrate robot rights, either now or in the future. In order to run the necessary demonstration and construct a robot that could qualify for meriting human-level rights (defined in terms of privileges, claims, powers, and/or immunities), one would need to build something that not only cannot give consent in advance of its own construction but which also could retroactively (after having been created) withdraw consent to its having been fabricated in the first place. So here's the paradox: The demonstration of a robot having rights might already violate the very rights that come to be demonstrated. Although one might be tempted to write this off as a kind of navel gazing exercise that is only of interest to philosophers concerned with logic puzzles, there is a real problem. Successfully demonstrating, for instance, that a robot is capable of feeling pain as a means by which to resolve questions concerning its moral or legal status runs the risk of violating the very thing it seeks to prove and demonstrate.[2]

### 3.3 Summary

Affirmative responses to the question "Can and should robots have rights?" tend to be future oriented and formulated in terms of a conditional statement. "If robots become sentient one day," as Neuhäuser (2015, 133) writes, "they will probably have to be granted moral claims. So far, however, such a situation is not yet in sight." Although robots are not capable of having rights at this particular point in time—when we can be reasonably certain that they are mere tools or instruments of human decision-making and action—that circumstance could change in the future; they could be (or eventually become) something more. This 'something more' is open to considerable discussion and debate as it concerns ontological properties or capabilities that are determined (in one way or another) to be the necessary and sufficient conditions for an entity to be granted both rights and responsibilities. And depending on who tells the story or makes the argument, these capabilities typically include things like: reason, consciousness, sentience, the ability to experience pleasure and/or pain, etc. But no matter what exact property or properties come to be identified as the qualifying criteria, the argument remains largely the same: once robots achieve this capability, then they can and should have rights.

This procedure is both sensible and problematic. It is sensible insofar it grounds decisions concerning moral status in what something actually is (or is not) as opposed to how it is experienced or appears to be. Appearances, as we have learned from Plato and the tradition of Platonism, change and are insubstantial. Being by contrast persists; it is real and substantive. In determining how something is to be treated on the basis of what it actually is—that is, deriving decisions concerning *ought* from *is*—we are (so it seems) grounding questions of moral standing in something that is real and true. But for all its advantages, this way of proceeding runs into considerable complications: (1) substantive problems concerning the identification of the exact property or properties that are considered to be the qualifying criteria, (2) terminological difficulties with defining these properties in the first place, (3) epistemological uncertainties with detecting the presence of the property in another entity, and (4) moral complications caused by the very effort to demonstrate of all of this. Deriving *ought* from *is* sounds reasonable, and it appears to be correct. But it is exceedingly difficult to deploy, maintain, and justify.

# 4  S1 !S2: Although Robots Can Have Rights, Robots Should Not Have Rights

In opposition to the previous approaches that derive an *ought* from an *is* (chapters 2 and 3), there are two other modalities that uphold (or at least seek to uphold) the is/ought distinction. In the first version, one affirms that robots can have rights but denies that this fact requires us to accord them social or moral standing. This is the argument that has been developed and defended by Joanna Bryson beginning with her provocatively titled essay "Robots Should Be Slaves" (2010). Bryson's argument goes like this: Robots are property. No matter how capable they are, appear to be, or may become; we are obligated not to be obligated by them. "It is," Bryson (2016a, 6) argues, "unquestionably within our society's capacity to define robots and other AI as moral agents and patients. In fact, many authors (both philosophers and technologists) are currently working on this project. It may be technically possible to create AI that would meet contemporary requirements for agency or patiency. But even if it is possible, neither of these two statements makes it either necessary or desirable that we should do so." In other words, it is entirely with in our power to define robots in such a way that they would have the rights of a moral or legal person, but we should not do so. Although this sounds pretty straight forward, there are complications with the argument that need to be addressed and taken into account.

## 4.1  The Argument

It is important, as Bryson explained to John Danaher in a 2017 podcast interview, to distinguish between two different kinds of artifacts—the robot and our legal/moral system. The question "Can robots have rights?" is not necessarily a matter of advancements in technology or with engineering practices. It could be, but then the "can question" would be (as it was in the previous chapter) a matter of some future achievement with technology. Bryson's focus is much more immediate and concerns not the technological but the legal/moral artifact. "The point is," as Bryson explained it, "that overnight a president could wave their hand and say by executive order that robots are now responsible agents … that could be done overnight. Law is like that" (Danaher and Bryson 2017). Consequently, the question "Can robots have rights?" is not exclusively a matter of advancements in engineering or technology—i.e., the creation of human-level artificial intelligence, or what Miller (2017) calls "maximally humanlike automata," possessing consciousness or whatever else is decided to be the qualifying property for moral and legal status. It may also be the result of a legal declaration. (And this is why, in the above quotation, Bryson deliberately uses the verb "define" rather than "create"). Robots *can* have rights insofar as someone in the position of legal or moral authority can decide to confer on them the status of moral and/or legal subjectivity. But, and this is Bryson's point, just because we *can* do this, does not mean that we should or that doing so would be a good idea.

The reason for this, Bryson argues, derives from the need to protect human individuals and social institutions. "My argument is this: given the inevitability of our ownership of robots, neglecting that they are essentially in our service would be unhealthy and inefficient. More importantly, it invites inappropriate decisions such as misassignations of responsibility or misappropriations of resources" (Bryson 2010, 65). This is why the word "slave," although somewhat harsh, is entirely appropriate.[1] Irrespective of what they are, what they can become, or what some users might assume them to be, we should treat all artifacts as mere tools and instruments. To its credit, this approach succeeds insofar as it reasserts and reconfirms the instrumental theory in the face of (perceived) challenges from a new kind of socially interactive and seemingly animate device. No matter how interactive, intelligent, or animated our robots become or appear to be, they should be, now and forever, considered to be instruments or "slaves" in our service, nothing more. "We design, manufacture, own and operate robots," Bryson (2010, 65) writes. "They are entirely our responsibility. We determine their goals and behaviour, either directly or indirectly through specifying their intelligence, or even more indirectly by specifying how they acquire their own intelligence. But at the end of every indirection lies the fact that there would be no robots on this planet if it weren't for deliberate human decisions to create them."

A similar argument is advanced by Roman V. Yampolskiy in the book *Artificial Superintelligence: A Futuristic Approach*:

> Finally I address a sub-branch of machine ethics that on the surface has little to do with safety but is claimed to play a role in decision making by ethical machines: robot rights (Roh 2009). RR [robot rights] asks if our mind children should be given rights, privileges and responsibilities enjoyed by those granted personhood by society. I believe the answer is a definite No. Although all humans are "created equal," machines should be inferior by design; they should have no rights and should be expendable as needed, making their use as tools much more beneficial for their creators. My viewpoint on this issue is easy to justify: Because machines cannot feel pain (Bishop 2009; Dennett 1978) (or less controversially can be designed not to feel anything), they cannot experience suffering if destroyed. The machines could certainly be our equals in ability but they should not be designed to be our equals in terms of rights. RR, if granted, would inevitably lead to civil rights, including voting rights. Given the predicted number of robots in the next few decades and the ease of copying potentially intelligent software, a society with voting artificially intelligent members will quickly become dominated by them. (Yampolskiy 2016, 140)

Similar to the argument offered by Bryson (but focusing on the technological and not legal/moral artifact), Yampolskiy argues that robots *could* be our equals in term of abilities, but they *should* not have rights. And they should not for two reasons. First, robots are and need to remain tools of human action, which are created and deployed for the benefit of their makers. Consequently, they should be designed and used as mere instruments to be exploited and disposed of as we see fit, and we should not design them to care about how they are treated or why. We have, in other words, an obligation not to build things to which we would feel obligated. Second, if we grant rights to robots, we are in danger of releasing a social transformation that we will not be able to control. For Yampolskiy this danger comes not in the form of the typical robot apocalypse (vividly imagined in science fiction) but in the rather less dramatic effect of voting rights and robot supported transformations in democratic governance. However it is accomplished, robot rights (or RR, as Yampolskiy writes it) would challenge or diminish human dignity and self-governance. Whether this is (or would be) the case, remains an open and debated question. Although Yampolskiy forcefully asserts his theory, he unfortunately offers little or no actual social/political data to support such a claim. It may sound intuitively correct (and perhaps this is due to the fact this scenario has been presented and prototyped in robot science fiction), but "sounding correct" is not sufficient proof.

Lantz Fleming Miller, who, as we have previously seen (chapter 2), excludes robots—even and especially maximally humanlike automata—from moral consideration on the basis of fundamental ontological differences, is sympathetic to these arguments. According to Miller's review of the existing literature, Bryson essentially gets it right but not necessarily for the right underlying reasons.

> Bryson (2000; 2010; see also Bryson and Kime 2011) contends automata remain machines thus tools of humans and it is (ontologically) incorrect to merge the two and ethically wrong to delude people by attempting to blur the distinction. This position is consistent with mine below, which, though, shows exactly which ontological distinction is critical in terms of rights. However, Bryson (2009) almost hits on this distinction in emphasizing it is humans who decide to design and construct machines and are thereby in a position to decide what kind of beings they are. (Miller 2015, 373)

Though Miller follows a procedure that derives different levels of moral status (either having full human rights or not) from different ontological conditions, his argument is consistent with Bryson's position[2]. Even if, Miller asserts, robots achieve humanlike abilities and are capable, in principle at least, of having human rights (or at least of leading us to ask the question concerning rights), we most certainly should not decide to extend rights to such artifacts. Despite their humanlike capabilities (or simulation of humanlike capabilities), they remain tools, and we are not obligated to extend to them human-level rights. "Humans are," Miller (2015, 387) concludes, "under no moral obligation to grant full human rights to entities possessing ontological properties critically different from them in terms of human rights bases."

## 4.2 Complications, Difficulties, and Potential Problems

These arguments are persuasive, precisely because they distinguish between a capability, i.e., "can have rights," and moral or legal imperatives, i.e., "should have rights." The importance of instituting and defending this distinction can be seen in the debates concerning the legal rights of corporations. According to both national and international laws, corporations are considered legal persons possessing a wide range of both responsibilities and rights. The limited liability corporation, in other words, can have (or perhaps stated more accurately, can be assigned) rights. In particular (and by way of example), the US Supreme Court recently ruled, in *Citizens United v. Federal Election Commission* 558 U.S. 310 (2010), that corporations have the right to free expression on par with that of a human person (Dowling and Miller 2014, 164). Despite this outcome, there has been and continues to be considerable debate as to whether this should in fact be the case. There are, as both critics and advocates recognize, significant moral and social consequences involved in extending such rights to the corporation. So even though it can and has been done (with the explicit support of legal institutions like the Supreme Court of the United States), that does not mean that it should be done or that doing so is necessarily just and right. Bryson, Mihailis Diamantis, and Thomas Grant (2017) advance a similar argument regarding the legal status of robots. "It is completely possible to declare a machine a legal person. The impulse to do so exists both at the individual level with academic proponents, and at the level of international governance with the European Parliament recommending consideration" (Bryson, Diamantis, and Grant 2017, 289). But just because this *can* be done, it does not mean that doing so would be a good idea or even desirable. In fact, "conferring legal personality on robots," as Bryson et al. (2017, 289) conclude, "is morally unnecessary and legally troublesome." There are, however, several difficulties with this way of proceeding, especially when it comes to dealing with robots.

### 4.2.1 Normative Proscriptions

First, these arguments involve or produce a kind of normative prohibition or even what one might call an asceticism.[3] Bryson's proposal in "Robots Should Be Slaves" issues what amounts to imperatives that take the form of social proscriptions directed to both designers and users. For designers, "Thou shalt not create robots to be companions." For users, no matter how interactive or capable a robot is (or can become), "Thou shalt not treat your robot as yourself." This is arguably a "conservative" strategy. "As I often recount," Bryson (2016b) explains in a blog post from June 23, 2016, "I got involved in AI ethics because I was dumbfounded that people attributed moral patiency to (thought I shouldn't unplug) Cog, the humanoid robot, when in fact it wasn't plugged in, and didn't work (this was 1993–1994). The processors of its brain couldn't talk to each other, they weren't properly grounded [earthed]. Just because it was shaped like a human and they'd watched Star Wars, passers-by thought it deserved more ethical consideration than they gave homeless people, who were actually people." Bryson's own involvement with AI ethics, as she tells the story, began as a result of observing other people (wrongly) attributing social standing to Cog, a device that was not even operational, and did not warrant or deserve a modicum of respect beyond that accorded to other "dumb" things. In response to this perceived problem, Bryson takes a conservative approach: She identifies a potentially dangerous error or mistake that could, according to her evaluation of things, produce all kinds of trouble, and in reaction to this issues imperatives designed to limit and even disarm this potentially risky and dangerous behavior before it ever happens.

This effort echoes a statement made by John McCarthy (1999, 15) over two decades ago: "It is also practically important to avoid making robots that are reasonable targets for either human sympathy or dislike. If robots are visibly sad, bored or angry, humans, starting with children, will react to them as persons. Then they would very likely come to occupy some status in human society. Human society is complicated enough already." McCarthy recognized that even if it were "possible to give robots human-like emotions," doing so would be a bad idea (McCarthy 1999, 15) and therefore should be deliberately avoided. Yampolskiy's (2016, 140) text issues similar imperatives that dictate what we should or should not do concerning robots, i.e., "they *should not* be designed to be our equals in terms of rights." And Matthias Scheutz (2012), as part of the "call to action" that concludes his analysis of "The Inherent Dangers of Unidirectional Emotional Bonds between Humans and Social Robots," codifies this in terms of a proposed regulatory policy:

> Another option might be, again, required by law, to make it part of a social robot's design, appearance, and behavior, that the robot continuously signal, unmistakably and clearly, to the human that it is a machine, that it does not have emotions, that it cannot reciprocate (very similar to the "smoking kills" labels on European cigarette packs). Of course, these reminders that robots are machines are no guarantee that people will not fall for them [just like the efficacy of the "smoking kills" notification remains questionable], but it might reduce the likelihood and extent to which people will form emotional bonds with robots. And it will present the challenge of walking a fine line between making interactions with robots easier and more natural, while clearly instilling in humans the belief that robots are human-made machines with no internal life (at least the present ones). (Scheutz 2012, 218)

These formulations are consistent with the fourth principle of the Engineering and Physical Sciences Research Council's (EPSRC) "Principles of Robotics: Regulating Robots in the Real World": "Robots are manufactured artefacts. They should not be designed in a deceptive way to exploit vulnerable users; instead their machine nature should be transparent" (Boden et al. 2011 and Boden et al. 2017, 127). Something similar appears in Miller (2015, 379): "We do not want any entities of type X constructed such that, if they were constructed, they would appear to merit human rights because they have some humanlike traits. They will not suffer if they are not constructed. If constructed, they would be of a different ontological type from us, and we would be under no moral obligation to let them join our society fully." And Schwitzgebel and Garza (2015, 113), who nevertheless take a different view of the future possibilities of robot rights (see chapter 2), issue similar advice: "Thus, we should generally try to avoid designing entities that don't deserve moral consideration but to which normal users are nonetheless inclined to give substantial moral consideration; and conversely, if we do some day create genuinely human-grade AIs who merit substantial moral concern, it would probably be good to design them so that they evoke the proper range of emotional responses from normal users. Maybe we can call this the Emotional Alignment Design Policy."

Admittedly most, if not all, normative ethics and systems of law work by issuing prohibitions and/or permissions—"Thou shalt not's" and "Thou shalt's." There is, therefore, nothing inherently incorrect or problematic about this particular procedure. A problem only occurs when such stipulations impose unrealistic or impractical restraints on behavior or strain against lived experience and developing social norms and conventions. And this appears to be the case with the normative proscriptions that have been issued by Bryson, Yampolskiy, Miller, and others. On the one hand, the attempt to limit innovation and regulate design may not be entirely workable in practice. As José Hernéndez-Orallo (2017, 448) points out, "Bryson (2010) and Yampolskiy (2016) contend that AI should never build anything with moral agency or patiency. However, this is easier said than done. On purpose or not, there will eventually be some AI systems that will approach any conventional limit set on a continuum. This actually sets a high responsibility on AI but also, once they are out there, on those who are in charge of evaluating and certifying that an agent has a particular psychometric profile." For Hernéndez-Orallo, the problem is not just with the practicalities of controlling who designs what, which is rather tricky to prescribe and regulate, but also with those individuals and organizations who would be charged with evaluating and testing applications either in advance of or just after having been released into the wild. Hernéndez-Orallo's point is simple: the level of responsibility that these proposals impose on both AI and its regulators would be "enormous" and practically untenable.

Erica Neely (2012 and 2014) offers a similar argument in her essay "Machines and the Moral Community":

> Joanna Bryson (2010) has argued that there is danger in being overly generous and extending rights to machines because we may waste energy and resources on entities which are undeserving of them; furthermore, this diverts our attention from the human problems which should be our concern. However, I believe she is too hasty in arguing that we simply can avoid the problem by not designing robots which deserve moral concern. While she is correct that we design and build robots, it should be clear to anyone who interacts with computers or software that we do not always correctly predict the results of our creations; moreover, there will almost certainly be someone who tries to design a self-aware autonomous machine simply because he can—because it would be interesting. As such, it is overly optimistic to believe we can simply avoid the question in the manner she suggests. (Neely 2014, 105)

Neely's point is that restricting innovation as a way of avoiding moral challenges and complications is simply not going to work. It is too hasty—dismissing the problem too quickly. The problem concerning the question of machine moral standing cannot, Neely argues, be quickly set aside by intentionally not doing something, especially something that has a high likelihood of actually happening.

On the other hand, "real world" (to reuse EPSRC's terminology) experience with robots introduces further complications. This includes not just anecdotal evidence gathered from the rather exceptional experiences of soldiers working with EOD robots on the battlefield (Singer 2009, Garreau 2007, Carpenter 2015) but numerous empirical studies of human/robot interaction (HRI) that verify the media equation initially proposed by Reeves and Nass (1996). Surveying a series of HRI studies with social robots, Scheutz (2012, 210) reports the following generalizable results: "that humans seem to prefer autonomous robots over nonautonomous robots when they have to work with them, that humans prefer human-like features (e.g. affect) in robots and that those features are correlated with beliefs about autonomy, and that a robot's presence can affect humans in a way that is usually only caused by the presence of another human." And in two recent studies by Rosenthal-von der Pütten et al. (2013) and Suzuki et al. (2015), researchers found that human users empathized with what appeared to be robot suffering, even when human test-subjects had prior experience with the device and knew that it was "just a machine." To put it in a rather crude vernacular form: Even when our head tells us "it's just a robot," our heart cannot help but feel for it. Consequently, despite the fact that "we don't want anyone risking their life because they mistakenly believe they are protecting more than the mindless Furby before them" (Schwitzgebel and Garza 2015, 113), this fact remains a theoretical ideal that is often not able to be realized in actual situations and circumstances. Informing users that "it is just a robot" (which is theoretically correct) fails to recognize or account for the actual data concerning the way human users respond to, interact with, and conceptualize these devices in practice. In other words (and to return to Bryson's AI ethics "conversion experience"), the fact that people responded to Cog as something more than just a dumb object is not a bug to be eliminated or fixed; it is a feature of human sociality.

### 4.2.2 Ethnocentrism

These normative proscriptions are made from and informed by a particular sociocultural position—specifically a Western European and predominantly Christian viewpoint—and therefore are not necessarily universally applicable or even widely agreed upon even at a theoretical level. Miller (2015, 374) notes this in a brief, self-reflective aside: "My approach, as will be seen, is hypothetical, resting conditionally upon a widely held belief about the nature of human rights and the properties of human beings that render them thusly deserving." As Miller recognizes, his argument is based on "a widely held belief." As such, it is not settled fact and is therefore exposed to challenges from other belief systems and competing hypotheses. Raya Jones makes this difference clear in her consideration of the work of Masahiro Mori, the Japanese robotics engineer who first formulated the uncanny valley hypothesis back in 1970. In a statement that directly contravenes Bryson's (2010, 71) "robot-as-slave" model, Mori (1981, 177; quoted in Jones 2016, 154) offers the following riposte: "There is no master-slave relationship between human beings and machines. The two are fused together in an interlocking entity." As Jones (2016, 154) explains, Mori's statement "connotes two ways that the concepts of 'human' and 'robot' can relate to each other. The 'master-slave' viewpoint that Mori eschews accords with individualism and the conventional understanding of technology in terms of its instrumentality. The viewpoint that Mori prompts is based in the Buddhist view of the interconnectedness of all things."[4]

The "Robots Should Be Slaves" hypothesis, then, is a statement that can only be made from, and in service to, a particular cultural norm. Things look entirely different from other sociocultural perspectives. Take, for example, the way that robots—actual existing robots—have been socially situated in Japanese culture, which is influenced and shaped by other traditions and practices. "On 7 November 2010," as Jennifer Robertson (2014, 590–91; also see Robertson 2017, 138) reports, the therapy robot "Paro was granted its own *koseki*, or household registry, from the mayor of Nanto City, Toyama Prefecture. Shibata Takanori, Paro's inventor, is listed as the robot's father (recalling the third of Tezuka's Ten Laws) and a 'birth date' of 17 September 2004 is recorded." The Japanese concept of *koseki*, as Robertson (2014, 578) characterizes it, is the traditional registry of the members of a household and by extension establishes proof of Japanese nationality or citizenship through patriarchal lineage. Although the *koseki* currently "carries no legal force," it still plays an important and influential role in Japanese society. For this reason, the act of granting Paro its own *koseki* was not just a savvy publicity stunt.

On the surface, the conferral of Paro's *koseki* may seem benign and inconsequential—even gimmicky. Quite the contrary. As I noted earlier, the *koseki* conflates family, nationality, and citizenship. It also legally and ideologically prioritizes the family (*ie*) over the individual as the fundamental social unit in Japanese society. Thus, a *zainichi* Korean man who was born, raised, and lives in Japan, who is married to a Japanese citizen, and whose natal family has lived in Japan for generations, can have neither his own *koseki* nor be included in the "family" portion of his wife's *koseki*; rather, his name is added to the "remarks" column of his wife's registry. By virtue of having a Japanese father, Paro is entitled to a *koseki*, which confirms the robot's Japanese citizenship." (Robertson 2014, 591 and Robertson 2017, 139)[5]

Related to this act has been "the granting of a *tokubetsu juminhyo* (special residency permit) between 2004 and 2012 to nine robots and dolls in localities throughout Japan" and to sixty-eight Japanese cartoon characters including Astro Boy and Doraemon (Robertson 2014, 592).

This is not to say that the proposals authored by Bryson (2014 and 2017), Miller (2015), Yampolskiy (2016), and EPSRC (2011 and 2017) are intentionally ethnocentric.[6] It simply points out that these arguments are formed and supported by cultural traditions, norms, and belief systems that are not necessarily universally valid or even widely accepted by others, and, what is perhaps worse, that this fact itself has rarely been recognize and identified as such. The EPSRC "Principles for Robotics," for example, confidently declares that "robots are simply tools of various kinds, albeit very special tools" (Boden 2017, 125). This seemingly correct and incontrovertible declaration, however, is issued from a distinctly European perspective and frame of reference—the individuals who were involved in the project had institutional homes at European universities and organizations, and EPSRC is a UK government agency. For this reason, the declaration and its significance cannot simply be taken as a matter of universal truth and settled fact. Although the document sought to devise realistic rules for real-world robots, the real world—especially when considered from a global perspective—is a more complicated and diverse place. The integration of robots into human society will need to learn to be sensitive and responsive to these important social and cultural differences.

### 4.2.3 Slavery 2.0

Finally, and perhaps most strikingly, there are potential problems and objections that accrue to the use of the term "slave" in these contexts. Bryson's (2010, 63) thesis "that robots should be built, marketed and considered legally as slaves" is not necessarily new or surprising; it is (or at least should be) rather familiar territory. It has been, as Lehman-Wilzig (1981, 449) pointed out, part and parcel of the robot's origin story and etymology from the time of Čapek's *R.U.R.*, and the concept continues to fuel much of the narrative conflicts in robot science fiction, from the replicants of *Do Androids Dream of Electric Sheep?* and its cinematic adaptations in *Bladerunner* and *Bladerunner 2049*, the Cylons of both iterations of *Battlestar Galactica* and its short-lived prequel *Caprica*, and the *hubots* of *Äkta människor* [*Real Humans*] or the synths of its English-language remake *Humans*. In fact, Isaac Asimov's three laws of robotics, which constitute what is perhaps the most direct and openly acknowledged point of contact between science fiction and actual work in robotics and robot ethics (McCauley 2007, Anderson 2008, Sloman 2010, Gunkel 2012), constitute what is arguably a code of conduct for artificial slaves. "To the extent that the three laws are effective," F. Patrick Hubbard (2011, 466) explains, "they function as a self-executing slave code for self-conscious robots: Do not harm masters;

obey masters except where harm to a master would result; and protect your owner's property interest in your wellbeing." Isiah Lavender takes this analysis one step further by acknowledging the fact that this concept of slavery, at least in the context of the United States, is inextricably connected to matters of race:

> Asimov himself proudly admits in his essay collection *Asimov on Science Fiction* (1981), "Robots can be the new servants—patient, uncomplaining, incapable of revolt. In human shape they can make use of the full range of technological tools devised for human beings and, when intelligent enough, can be friends as well as servants" (88–89). Asimov's robots resonate with the antebellum South's myth of a happy darkie—a primitive, childlike worker without a soul, incapable of much thought—cared for by the benevolent and wise master. This resonance is hard to ignore. As Edward James [1990, 40] has observed in his own critique of Asimov, "The Three Laws restrain robots, just as the slave owner expected (or hoped) that his black slaves would be restrained by custom, fear and conditioning to obey his every order." (Lavender 2011, 62)

But the idea that robots are and should be slaves is not something that is limited to contemporary science fiction. On the one hand, as Kevin LaGrandeur (2013) demonstrates, there are numerous premodern formulations in both literature and natural philosophy. "The promise and peril of artificial, intelligent servants," LaGrandeur (2013, 9) explains, "was first implicitly laid out over 2000 years ago by Aristotle." Although a type of artificial servant had been previously depicted in Homer's *Iliad* with the tripods of Hephaestus [Hēphaistos in transliterated Greek] "that can roll themselves in and out of the gods' banquet room" (LaGrandeur 2013, 9), it was Aristotle who was the first "to actually discuss their uses and advantages" (LaGrandeur 2013, 9):

> In Book 4 of his *Politics*, he [Aristotle] refers back to Hephaistos' intelligent artifacts and argues that "if, in like manner, the shuttle would weave and the plectrum touch the lyre, chief workmen would not want servants nor masters slaves" (1253b38-1254a1). The advantages of intelligent, artificial servants are clear in this statement. They allow for work to be done for their owners with no intermediation. Also they allow for a person to forgo both the ethical problems of owning human slaves (which, as Aristotle notes, is not approved by all Athenians) and the possible dangers and aggravations as well (1253b20-23, and 1255b12-15). (LaGrandeur 2013, 9)

Aristotle, therefore, accurately described robotic slaves *avant la lettre*. The "intelligent, artificial servants" that he imagined would not only work tirelessly on our behalf, but would, precisely because of this, make human servitude and bondage virtually unnecessary. Since the time of Aristotle, many versions of what LaGrandeur calls "artificial slaves" appear in ancient, medieval, and Renaissance sources. Consequently, the dream of robot slaves is old and rooted in some rather ancient ideas and prototypes.

On the other hand, postwar predictions about the eventual implementation of robots in both industry and the home draw on and mobilize the same formula. As early as 1950, Norbert Wiener, the progenitor of the science of cybernetics, suggested that: "the automatic machine, whatever we may think of any feelings it may have or may not have, is the precise economic equivalent of slave labor" (Wiener 1954, 162; also Wiener 1996, 27). In the January 1957 issue of *Mechanix Illustrated*, a popular science and technology magazine published in the United States, one finds a story titled "You will Own 'Slaves' by 1965." The article begins with the following rather insensitive characterization of the opportunities of robot servitude: "In 1863, Abe Lincoln freed the slaves. But by 1965, slavery will be back! We'll all have personal slaves again, only this time we won't fight a Civil War over them. Slavery will be here to stay. Don't be alarmed. We mean robot 'slaves'" (Binder 1957, 62). And in 1988, two legal scholars, Sohail Inayatullah and Phil McNally (1988, 131), argued that slavery would be the most reasonable way to accommodate and respond to the opportunities and challenges posed by innovations in robotics: "Given the structure of dominance in the world today: between nations, peoples, races, and sexes, the most likely body of legal theory that will be applied to robots will be that which sees robots as slaves. They will be ours to use and abuse."

Consequently, Bryson's (2010, 63) thesis "that robots should be built, marketed and considered legally as slaves," despite (or perhaps because of) the complicated history that surrounds the term "slave," has considerable traction. "Modern robots," as LaGrandeur (2013, 161) summarizes it, "are chiefly meant to perform the same jobs as slaves—jobs that are dirty, dangerous, or monotonous—thereby freeing their owners to pursue more lofty and comfortable pursuits." And contemporary efforts to address real-world dilemmas concerning the legal status of robots are rediscovering useful resources in already existing legal systems concerning slaves and slavery. Take for example the following advice provided by Luciano Floridi in an editorial published in *Philosophy & Technology*:

There is no need to adopt science fiction solutions to solve practical problems of legal liability with which jurisprudence has been dealing successfully for a long time. If robots become one day as good as human agents—think of Droids in *Star Wars*—we may adapt rules as old as Roman law, according to which the owner of an enslaved person was responsible for any damage caused by that person (*respondeat superior*). As the Romans already knew, attributing some kind of legal personality to robots would deresponsabilise those who should control them. Not to speak of the counterintuitive attribution of rights." (Floridi 2017, 4)

The possibility of adopting and repurposing ancient Roman laws regarding slaves to resolve contemporary questions concerning legal liability with autonomous technology is something that is already gaining traction in the new field of robot law (Pagallo 2013 and Ashrafian 2015b). Recently a number of legal scholar have argued in favor of the robot-as-slave model as a way to contend with the contractual and liability issues regarding software bots (Schweighofer 2001, 45–53; John 2007, 91; Günther 2016, 51–52). Reviewing this literature, Leroux et al. (2012, 60) provide the following advice concerning the evolving legal status of Softwareagents and its possible extension to embodied robots:

> Softwareagents as Electronic Slaves: Another approach could be to see such Softwareagents as objects with limited legal capacity following the law of *slave in ius civile*. In Roman civil law, slaves had no capacity to bear rights and obligations while acting under his own name. They had no legal right to enter into contract on their own behalf but were able to enter into contract as agent for their master. Thus, the actions of a slave could be attributed to the master. This same logic could apply to the consideration of robots because slaves and robots are similar in two main respects: they both have a limited capability of legal acting and rights or obligations while acting in their own name.

So it looks as if the robot-as-slave model not only works for the purpose of generating compelling fictional narratives but could also be put to work for responding to the questions concerning responsibility and rights with robots insofar as there are already bodies of law that could be adopted and redeployed to cover the opportunities and challenges presented by an increasing number of autonomous machines in contemporary society. But not so fast; things may be more complicated than they initially appear to be.

**Terminology** When mobilizing the robot-as-slave model, everything depends on how we understand and operationalize the term "slave." According to Aristotle (1944, 1253b26–40), slaves are a specific kind or species of tool:

Of tools [ὄργανον] some are lifeless [ἄψυχα] and others living [ἔμψυχα] (for example, for a helmsman the rudder is a lifeless tool and the look-out man a live tool—for an assistant in the arts belongs to the class of tools), so also an article of property is a tool for the purpose of life, and property generally is a collection of tools, and a slave [δοῦλος] is a live article of property. And every assistant [ὑπηρετῶν] is as it were a tool that serves for several tools; for if every tool could perform its own work when ordered, or by seeing what to do in advance, like the statues of Daedalus in the story, or the tripods of Hephaestus which the poet says "enter self-moved the company divine,"—if thus shuttles wove and quills played harps of themselves, master-craftsmen would have no need of assistants and masters no need of slaves.

Formulated in this way, there are two categories of tools, living and nonliving, and slaves (along with "assistants" or ὑπηρετῶν) are living tools. This means, on the one hand, and in conformity with the instrumental theory of technology, that actions undertaking by means of a tool—living or nonliving—are ultimately the responsibility of the user/owner. Or, as it was codified in Jewish law, "*yad eved k'yad rabbo*—the hand of the slave is like the hand of its master" (Lehman-Wilzig 1981, 449). And, on the other hand, a tool is just an object—a mere thing or "an article of property" (Aristotle's formulation)—that can be used and/or abused as the owner of the instrument decides and sees fit.[7] This was the case, as reported by Leopold (1966, 237), with Odysseus, who disposed of his slave girls upon returning to Ithaca. This understanding of the term "slave," however, is limited and not entirely accurate, especially when considered from the perspective of Roman law to which Floridi and several others appeal.

Roman law did not institute a simple dichotomy between citizens and slaves but provided for differing degrees of social standing for different categories of things. "Under the *Ius Gentium* law," as Hutan Ashrafian (2015b, 324) points out, "Roman citizens were given a full complement of rights (through *Ius Civile*) whilst there were several classes of free individuals, including people of Latin (from *Latium*), *Peregrinus* (Provincial people from throughout the empire) and *Libertus* (Freed slave) status." Furthermore, even those individuals on the low end of the spectrum, who were regarded as chattel, had some limited rights especially in matters having to do with contracts. "While Roman lawyers," Pagallo (2013, 103) points out, "invented forms of agency and autonomy for mere things without legal personality, their aim was to strike a balance between the interest of the masters not to be negatively affected by the business of their slaves and the claim of the slaves' counterparts to be able to safely interact or do business with them." For these reasons, both Pagallo (2013, 102–104) and Ashrafian (2015b) argue that robots, instead of being regarded as mere tools, could come to occupy a number of different positions within the Roman social spectrum and that deciding where robots stood in regards to these existing categories would require some critical decision-making and debate. "Here," Ashrafian (2015b, 325) concludes, "robots would likely occupy *Peregrinus* or possibly partial-Latin status, where they would not self-replicate, stand in public office or own land and business but would be protected by the law and have the ability to contribute to society through examples such as defending nations and participating in the healthcare sector." Complicating the issue is the fact that Roman law, like all legal systems, was not static but evolved and changed over time.

> After some time, the Romans introduced the Edict of Caracalla or Antonine Constitution (*Constitutio Antoniniana*) in 212 AD (likely to increase the number of individuals subject to taxation). Here Roman citizenship was granted to all "freeborn" men throughout the Empire whereas all freeborn women in the empire would receive the same rights as Roman women. Taken to its eventual conclusion, the continual advances in artificially intelligence agents and robots may herald their status of fully-fledged personalities with an accompanying level of higher legal and moral responsibilities but also a higher degree of rights … Consequently one possibility is that a legal personhood status might ensue for robots and artificial intelligence agents as result of a future "Caracalla approach." (Ashrafian 2015b, 325)

That future has arrived, with Caracalla-type proposals already having been advanced and put to vote. In a draft report to the European Parliament's Commission on Civil Law Rules on Robotics, the Committee on Legal Affairs proposed "creating a specific legal status for robots, so that at least the most sophisticated autonomous robots could be established as having the status of electronic persons with specific rights and obligations" (Committee on Legal Affairs 2016, 12) including taxation and the payment of social security contributions (Committee on Legal Affairs 2016, 10). As Floridi (2017, 3) explains, the reasoning behind this proposal is rather intuitive:

> Robots replace human workers. Retraining unemployed people was never easy, but it is more challenging now that technological disruption is spreading so rapidly, widely, and unpredictably. Today, a bus driver replaced by a driverless bus is unlikely to become a web master, not least because even that job is at risk of automation. There will be many new forms of employment in other corners of the infosphere. Think of how many people have opened virtual shops on eBay. But these will require new

and different skills. So more education and a universal basic income may be needed to mitigate the impact of robotics on the labour market, while ensuring a more equitable redistribution of its economic benefits. This means that society will need more resources. Unfortunately, robots do not pay taxes. And it is unlikely that more profitable companies may pay enough more taxes to compensate for the loss of revenues. So robots cause a higher demand for taxpayers' money but also a lower supply of it.

Robots will, for better or worse, have—and are already having—an effect on employment opportunity, and this "technological unemployment" (Keynes 1963) will affect not only individual workers, but also national economies and social structures. The solution that was proposed by the draft document takes the form of what many have called a "robot tax." Critical opposition to this idea was vehement, and, for this reason, the final version of the proposal backed-off from this stipulation. Despite these qualifications, however, what we see in this document (which is, it should be remembered, not law but a call for the creation of law), is movement in the direction of what Ashrafian (2015b, 325) called "a Caracalla approach"—the granting of special social status to previously excluded entities for the purpose of expanding taxation opportunities. What this means is that even if we are persuaded by and accept the "model of robot-as-slave" (Bryson 2010, 71) or what Lehman-Wilzig (1981, 449) calls "the robot-slave legal parallelism," *slave* remains a complex legal category that is not completely deprived of, but already granted some level of, legal rights and responsibilities.

**Blowback** The robot-as-slave model, which was initially advanced in order to tether machine action to the human owner of the device and effectively dismiss talk of "robot rights," might actually reintroduce the question of rights, albeit from the negative perspective of punishment. If we attend to the details—if we are, following Bryson (2010, 70) truly interested in "getting the metaphor right"—then we have to recognize that the slave has never simply been a neutral tool that is used by and always under the complete control of its master. In many of the legal traditions—Roman, Jewish, and US law (see Lehman-Wilzig 1981)—there are situations where the slave and not the master, or *dominus*, is to be held liable for wrongdoing and thereby becomes subject to legal action and punishment. As Lehman-Wilzig (1981, 449) points out by way of W. W. Buckland's study of *The Roman Law of Slavery* (1970), "Roman law considered the slave in a different light: 'a noxal action lies against the *dominus*, under which he must pay the damages ordinarily due for such a wrong, or hand over the slave to the injured person.' However, the Roman 'system of noxal action applies … to cases of civil injury, involving a liability to money damages; it does not apply to … criminal proceedings of any kind.' And even in civil cases, the master was free from personal liability if there existed a total absence of complicity." So if we succeed in "getting the metaphor right," there are two related difficulties that complicate the image.

On the one hand, the existing legal frameworks for slavery require the possibility of punishing the slave in situations (predominantly, but not exclusively, those involving some form of criminal wrongdoing) where the master is released from culpability. Because "retributive punishment" is, as Danaher (2016) points out, a crucial component of most if not all legal systems, one needs to be able to identify who or what can be punished for legal transgressions. But what can it mean to punish a robot? As Peter Asaro (2012, 182) has explained, this question is as complex as it is difficult to resolve:

> Robots do have bodies to kick though it is not clear that kicking them would achieve the traditional goals of punishment. The various forms of corporal punishment presuppose additional desires and fears of being human that may not readily apply to robots—pain, freedom of movement, mortality, etc. Thus, torture, imprisonment and destruction are not likely to be effective in achieving justice, reform or deterrence in robots. There may be a policy to destroy any robots that do harm but, as is the case with animals that harm people, it would be a preventative measure to avoid future harms rather than a true punishment. Whether it might be possible to build in a technological means to enable genuine punishment in robots is an open question.

Punishment is only possible on the prior condition of there being something that would matter to the entity being punished, e.g., its own welfare, its continued existence, its right to be free from pain, etc. When, as Danaher (2016, 306) recounts, the protagonist of the British comedy *Fawlty Towers* (Basil Fawlty, played by the actor John Cleese) beats his broken-down automobile with a stick, the car does not care, and it is this "not caring" that makes the act impotent and comical. Punishment only matters—only makes a difference—when it affects something that matters for the entity that is subject to punishment. But this capability—which is inherent to and necessary for the robot to be a slave—immediately complicates the strict instrumentalist view that was supposed to be reasserted by way of the robot-as-slave model. Consequently, the legal realities of slavery make application of the slave metaphor to robots more complicated, if not contradictory.

On the other hand, and following from this, if an entity can in fact be subject to punishment, then it has, at least *via negativa*, certain rights that will need to be taken into account if only for the purposes of formulating and executing a punishment that would matter and be effective. Incarceration, for instance, is efficacious only on the prior condition that an individual has the presumed right to free and unhindered movement (which is something that can be formally articulated in terms of several, if not all, of the Hohfeldian incidents, i.e., it could be described as a privilege granted individuals, a claim one has to free and unhindered movement, a power concerning the ability to exercise such movement, or even an immunity from undue restrictions). If robots are (or at least are considered to be) slaves, and slaves by definition can be subject to some form of retribution, then the robot-slave will need to have some legally recognized right or rights—no matter how minimal—that constitute the condition for the possibility of its punishment. So in the end, the robot-as-slave model—which sought to justify not just the withholding of rights from robots but the exclusion of the question from serious consideration—might actually need to consider and mobilize the very thing it sought to set aside. If "robots should be built, marketed and considered legally as slaves" (Bryson 2010, 63) and if we are in fact committed to "getting the metaphor right" (Bryson 2010, 70), this is only possible on the prior condition that robots can have some level of rights and interests.

**Social Costs** Finally, even if we disregard or temporarily bracket these complications accruing to the term "slave," there are significant social costs with this "slavery 2.0" proposal. But again we should be careful to get the metaphor right. The problem here, as Brooks (2002) insightfully points out, is not with the manufacture and use of machines that will work for us on our behalf. We should not, in other words, get hung up on mere words, like "slave." The problem has to do with the different kinds of robotic mechanisms to which this term might come to be applied:

Fortunately we are not doomed to create a race of slaves that is unethical to have as slaves. Our refrigerators work twenty-four hours a day seven days a week, and we do not feel the slightest moral concern for them. We will make many robots that are equally unemotional, unconscious, and unempathetic. We will use them as slaves just as we use our dishwashers, vacuum cleaners, and automobiles today. But those that we make more intelligent, that we give emotions to, and that we empathize with, will be a problem. We had better be careful just what we build, because we might end up liking them, and then we will be morally responsible for their well-being. Sort of like children. (Brooks 2002, 195)

The problem with robotic slavery—when it is cited as a problem—is more often than not assumed to be what the enslaved entity might feel or want, assuming that it would be possible for it to "feel" or "want" anything. This issue, which is given considerable play in science fiction from *R.U.R.* to *Bladerunner* and from *Robopocalypse* to *Battlestar Galactica* and *Humans*, is something that concerns Levy (2009, 212):

Within a few decades robots will be in almost every home, cooking, cleaning, doing our drudge work. But what will happen if they evolve to such an extent that they do not actually want to do our drudge work? Do we have any right to enslave them simply because they are not human? Is it fair and reasonable to deprive them of an existence full of pleasure and relaxation? Are we able to program a robot to have a soul and, if so, should we have the right to exercise influence and control over that soul? Even worse, if our robots have souls, do we have the right to switch off their souls if the mood takes us, or is that murder? If robots have consciousness, is it reasonable for us to argue that, because we gave them their ability to think for themselves, we should be able to command them to do our bidding, to enslave them? The answer, surely, should be "no," for the same moral reasons that we ought not enslave our children even though they owe us their very existence and their ability to think. And if robots are free to lead "normal" lives, whatever "normal" will come to mean for robot citizens, will they be able to claim social benefits, or free medical care and education, or unemployment benefits?

For Levy, the problem has to do with what the robot might want or feel and is therefore predicated upon the achievement of certain moral/emotional capabilities that for now remain a future possibility. According to Levy then, the important issue is a question about the rights of the robot for the sake of the robot. Or, as in some other cases (like many of the science fiction scenarios), the sake of prophylactically appeasing the robots so that they do not rise up and kill us. But this is precisely what Bryson targets and argues is a straw man. "The people that are horrified by the idea of robots being trapped in servitude or ownership," Bryson explains (Danaher and Bryson 2017), "are the people that are identifying with them too strongly." There is, as she argues, no reason for us to design robots that would either have or invite this problem, and, what is worse, it would be wrong (or at least morally problematic) for us to create such a mechanism in the first place. We can, she suggests, manufacture what are ostensibly mindless slaves—robot servants that, like our refrigerator and the other technological devices in our lives, work tireless for us, that do not mind doing so, and that are clearly identified to us as such so that no one ever get confused or makes the mistake of misattribution.

But this might be missing the point. The problem with "robotic slavery" (if that is indeed the term we want to use, and there are, as we have already seen, good reasons to hesitate[8]) is not necessarily with what one might think, namely, how the robot-slave might feel about its subjugation, bondage, or limited rights. This is, in fact, an argument that is based on a kind of speculation that simply conflates robot servitude with human slavery—something that is both morally questionable and that Bryson is careful to avoid. The problem is on the side of the master and the effect institutionalized slavery has on human individuals and communities. In *The Phenomenology of Spirit,* G. W. F. Hegel (1977, 111–119) famously demonstrated that slavery has negative consequences for the master who is, due to the very logic of the master/slave dialectic, incapable of achieving independence insofar as he is and remains beholden to the work performed by the slave. This philosophical insight has been verified by observational data. As Alexis de Tocqueville (1899, 361) reported during his travels in the southern United States, slavery was not just a problem for the slave, who obviously suffered under the yoke of forced labor and prejudice; it also had deleterious effects on the master and his social institutions. "Servitude, which debases the slave, impoverishes the master" (de Tocqueville 1899, 361). The full impact of this "all-pervading corruption produced by slavery" (Jacobs 2001, 44) is perhaps best described through the first-person account recorded by Harriet Ann Jacobs in *Incidents in the Life of a Slave Girl* (2001, 46): "I can testify, from my own experience and observation, that slavery is a curse to the whites as well as to the blacks. It makes the white fathers cruel and sensual; the sons violent and licentious; it contaminates the daughters, and makes the wives wretched." Clearly Bryson's use of the term "slave" is provocative and morally charged, and it would be impetuous to presume that her proposal for a kind of "slavery 2.0" would be the same or even substantially similar to what had occurred (and is still unfortunately occurring) with human bondage. But, and by the same token, we should also not dismiss or fail to take into account the documented evidence and historical data concerning slave-owning societies and how institutionalized forms of slavery affected both individuals and human communities.

Miller (2017) has similar concerns. Although he finds Bryson's general outlook to have "merit in spirit," it is her specific choice of terminology that is troubling. "The problem," Miller (2017, 5) argues, "is the term 'slave.' If slavery is, as most of the world now concurs, not morally good, it is reasonable to deduce that not only should no one be anyone's slave, but also no one should be anyone's master. There is something about the relationship that is wrong." For Miller, the problem with slavery is—echoing the arguments provided by Hegel, de Tocqueville, and Jacobs—the institution of bondage itself and the way that that particular relationship diminishes both participants, especially the one who would occupy the position of master:

> There is some harm to one's own higher moral values and moral character if one establishes oneself as master. One may propose that one value that leads one to be master is excessive ease and comfort. Even Locke, notoriously a plantation owner, at the least observed (if hypocritically) that there is a rightful amount of goods one may attain to maintain subsistence, and beyond that is taking from others. Surely, some persons, such as generals or corporate leaders, who cannot manage all their personal affairs alone, need personal assistance. Yet, such tasks can be handled by paid—sufficiently paid—voluntary servants. One need not have absolute control over such a life to get the tasks done, and ethically so. The problem of using and treating machines as slaves is that one perpetuates a value that sustains the inappropriate agent character, seeing the world and its denizens as one's slaves. You simply should not treat the world as a place in which your will is absolute. You thereby only strengthen that absolutist, disregarding will. (Miller 2017, 5)

A good, albeit somewhat extreme, illustration of this problem can be found in the literature on sex robots. "To state the obvious," as Charles Ess (2016, 65) explains, "attention (especially male attention) to developing robots that might satisfy our erotic and sexual desires and interests has a long, long history." That history, as John Sullins (2012, 398) explains, goes at least as far back as the story of Pygmalion, which is recounted in Ovid's *Metamorphosis*. "The Pygmalion theme," as Ess (2016, 65) points out, "continues through various versions of fembots, sexbots, female AIs, etc. that have populated movies from *Metropolis* to *I, Robot* to *Ex Machina*: a female robot that can serve as one's lover, if not sex slave." The idea of the sex robot—which is arguably still more of an idea that is under development insofar as "there is no working definition of a sex robot, and in reality there are really no sex robots" (Richardson 2016a, 48)[9]—has ignited considerable controversy and criticism.

The target of these efforts, however, is generally not located on the side of the robot and how it might "feel" about being pressed into service as a sex worker. It seems entirely possible that sex robots, following the procedures and protocols advanced by Bryson and others, could simply be designed not to care about how and why they are used for sexual gratification or (perhaps better) to be programmed to "like" servicing the carnal desires of their human users. The problem then is not with the robot sex-worker; the problem is on the side of the human user. Reviewing the available literature on the subject, Noel Sharkey et al. (2017, 22) provide the following inventory:

> The scholars cited here are pretty much in agreement that sex with robots could or will lead to social isolation [Whitby 2011]. The reasons given varied: spending time in a robot relationship could create an inability to form human friendships [Sullins 2012]; robot don't meet the species specific needs of humans [Richardson 2016]; sex robots could desensitize humans to intimacy and empathy, which can only be developed through experiencing human interaction and mutual consenting relationships [Kaye 2016, Vallor 2015]; real sexual relationships could become overwhelming because relations with robots are easier [Turkle 2012].

But the "victims" of robot sex are not just users, and the problems are not limited to the potential isolation, addiction, social de-skilling, and emotional desensitization they could suffer. The deleterious effects extend to others and could have an adverse impact across the social spectrum.[10] Kathleen Richardson's "Campaign Against Sex Robots" (initially introduced in the course of a paper presented at the Ethicomp conference in 2015) is not concerned with the welfare of the robotic sex worker (Richardson is certainly no "robot liberation activist") nor the social/emotional problems that might come to be experienced by individuals who engage in the risky behavior of robot sex. Her concern is: (1) the way that the rhetoric concerning sex robots already mobilizes and draws on disturbing parallels with human prostitution; and (2) how the continued use of this comparison in the literature does not, as Levy and others have suggested, solve the problems of human sex trafficking and prostitution (which as Richardson correctly points out are the most common forms of real world slavery in the twenty-first century), but continues the objectification of women and girls that has been and remains the direct consequence of the commodification of sex.[11] "Over the last decades," the website for the Campaign Against Sex Robots (2016) explains, "an increasing effort from both academia and industry has gone into the development of sex robots—that is, machines in the form of women or children for use as sex objects, substitutes for human partners or prostituted persons. The Campaign Against Sex Robots highlights that these kinds of robots are potentially harmful and will contribute to inequalities in society."

## 4.3 Summary

The recommendation that "robots should be slaves" has been derived from and justified by recognizing that it is entirely possible to make robots that would have rights—either by way of future advancements in technology (Miller) or here and now by way of decree (Bryson)—but that we should not do so. In other words, just because we *can* do something, does not mean that we *should*, especially if doing so could have deleterious effects on human individuals and institutions. For this reason, it is argued that robots ought to be situated in the social position that had been occupied (in previous eras) by slaves or servants. As such, robots will be a kind of "animate tool" (to use the Aristotelian formulation) that we own and that can be employed as human users decide and see fit. This decision, which ultimately keeps a human in the loop for dealing with matters of both responsibility and rights, is not only supported by the is/ought distinction that had been developed by David Hume but is aligned with the exigencies of law. As Bert-Jaap Koops (2008, 158) explains: "The reduction of *ought* or *ought not* to *can* or *cannot* threatens the flexibility and human interpretation of norms that are fundamental elements of law in practice." In other words, declaring that "robots should be slaves"—or "servants" (if one prefers to use a less polarizing term)—seems to be a reasonable and practical response to the question concerning robots and rights.

As reasonable and practical as the proposal initially sounds, however, there are considerable problems that complicate the picture. First, this way of thinking institutes social prohibitions that are not only unrealistic, when considered from the side of device designers and manufacturers, but are also actively challenged by the moral intuitions and practical experiences of users, who are already responding to robots—and not just friendly seeming sociable robots but even apparently cold industrial mechanisms—as something more than a mere instrument, something possessing a social presence that does have an effect and impact on us. Second, the robot-as-slave model is uncritically ethnocentric. It is a proposal that is: made from a distinct sociocultural perspective; informed by specific moral, legal, and religious/philosophical traditions; and elevated to the position of a universal truth without recognition of this fact or its consequences. Other cultures and traditions approach robots differently and therefore find the robot-as-slave model to be not only out of sync with their customs and experiences but a kind of cultural imperialism dressed-up and disguised as science fact. Finally, the concept of slavery or robotic servitude (which, it should be recalled, is inescapably part of the etymological legacy of the word "robota") already has social and political consequences that make its use problematic. In other words, even if we succeed in "getting the metaphor right" (Bryson 2010, 70), it may be the wrong metaphor to begin with insofar as: a) slaves have never been mere tools but have had some access to rights, if only for the purposes of taxation or retributive punishment; and b) the corrupting influence of the institution of slavery concerns not just the enslaved population but those who would be in the position of mastery.

# 5  !S1 S2: Even If Robots Cannot Have Rights, Robots Should Have Rights

The final modality also supports the independence and asymmetry of the two statements, but it does so by denying the first and affirming the second. In this case, which is something proposed and developed by Kate Darling (2012 and 2016a), robots, at least in term of the currently available technology, cannot have rights. They do not, at least at this particular point in time, possess the necessary capabilities or properties to be considered full moral and legal subjects. Despite this fact, there is, Darling asserts, something qualitatively different about the way we encounter and perceive robots, especially social robots. "Looking at state of the art technology, our robots are nowhere close to the intelligence and complexity of humans or animals, nor will they reach this stage in the near future. And yet, while it seems far-fetched for a robot's legal status to differ from that of a toaster,[1] there is already a notable difference in how we interact with certain types of robotic objects" (Darling 2012, 1). This occurs, Darling continues, principally due to our tendencies to anthropomorphize things by projecting onto them cognitive capabilities, emotions, and motivations that do not necessarily exist. Socially interactive robots, in particular, are intentionally designed to leverage and manipulate this proclivity. "Social robots," Darling (2012, 1) explains, "play off of this tendency by mimicking cues that we automatically associate with certain states of mind or feelings. Even in today's primitive form, this can elicit emotional reactions from people that are similar, for instance, to how we react to animals and to each other." And it is this emotional reaction, Darling argues, that necessitates obligations in the face of social robots. "Given that many people already feel strongly about state-of-the-art social robot 'abuse,' it may soon become more widely perceived as out of line with our social values to treat robotic companions in a way that we would not treat our pets" (Darling 2012, 1).

## 5.1 Arguments and Evidence

This argument proceeds (even if Darling does not explicitly identify it as such) according to an interest theory of rights. Although robots cannot, at least not at this particular point in time, have rights or achieve levels of intelligence and complexity that make a claim to rights both possible and necessary, there is an overwhelming interest in extending to them some level of recognition and protection. As proof of this claim, Darling offer anecdotal evidence, results from a workshop demonstration, and at least one laboratory experiment with human subjects.

### 5.1.1 Anecdotes and Stories

Darling recounts (and often begins her public presentations with) the tragic story of Hitchbot. Hitchbot was a hitchhiking robot, created by David Harris Smith and Frauke Zeller, that successfully made its way across Canada and Europe only to be brutally vandalized at the beginning of a similar effort to cross the United States. For Darling, what is important in this story, and what she highlights in her work, is the way that people responded to Hitchbot's demise. "Honestly I was a little surprised that it took this long for something bad to happen to Hitchbot. But I was even more surprised by the amount of attention that this case got. I mean it made international headlines and there was an outpouring of sympathy and support from thousands and thousands of people for Hitchbot" (Darling 2015). To illustrate this, Darling reviews tweets from individuals who not only expressed a sense of loss over the "death of Hitchbot"—but took it upon themselves to apologize directly to the robot for the cruelty that had been visited upon it by cold and insensitive humans. At other times, Darling has referenced and described the experiences people have with their Roomba vacuum cleaners—the fact that they give them names and "feel bad for the vacuum cleaner when it gets stuck under the couch" (Darling 2016b). And in another "more extreme example of this" (Darling 2016b), she recounts the experience of US soldiers with military robots, referencing the work done by Julie Carpenter (2015): "There are countless stories of soldiers becoming emotionally attached to the robots that they work with. So they'll give them names, they'll give them medals of honor, they'll have funerals for them when they die. And the interesting thing about these robots is that they're not designed to evoke this at all; they're just sticks on wheels" (Darling

2016b). Although it is not explicitly acknowledged as such, all three anecdotes (and there are several more involving other robotic objects) exemplify what Reeves and Nass (1996) call "the media effect"—the fact that human beings will accord social standing to mechanisms that exhibit some level of social presence (even a very minimal level, like the independent movement of the Roomba), whether this is intentionally designed into the artifact or simply perceived by the human user/observer of the device.

But Darling's efforts are not limited to recounting stories provided by others; she has also sought to test and verify these insights by way of her own research. In 2013, Darling along with Hannes Gassert, ran a workshop at LIFT13, an academic conference in Geneva, Switzerland. The workshop, which had a title that called upon and utilized Hume's is/ought distinction—"Harming and Protecting Robots: Can We, Should We?"—sought to "test whether people felt differently about social robots than they would everyday objects, like toasters" (Darling and Hauert 2013).

In the workshop, groups of participants [four groups of six for a total of twenty-four] were given Pleos—cute robotic dinosaurs that are roughly the size of small cats. After interacting with the robots and performing various tasks with them, the groups were asked to tie up, strike, and "kill" their Pleos. Drama ensued, with many of the participants refusing to "hurt" the robots, and even physically protecting them from being struck by fellow group members ... While everyone in the room was fully aware that the robot was just simulating its pain, most participants giggled nervously and felt a distinct sense of discomfort when it whimpered while it was being broken. (Darling 2016a, 222)

Although the demonstration was conducted "in a non-scientific setting" (Darling and Vedantam 2017), meaning that the results obtained are anecdotal at best (more on this in a moment), it did provide a compelling illustration of something that has been previously tested and verified in laboratory studies involving robot abuse[2] undertaken by Christopher Bartnek and Jun Hu (2008) and Astrid M. Rosenthal-von der Pütten et al. (2013). The former utilized both a Lego robot and Crawling Microbug Robots; the latter employed Pleo dinosaur toys, like Darling's workshop.

### 5.1.2 Scientific Studies

The one scientific study that has been performed and published by Darling, involves a Milgram-like obedience experiment designed to test the effect of framing on human empathy. In this laboratory experiment, which leverages the basic approach of Bartnek and Hu's (2008) second robot abuse study with the Crawling Microbug robot, Darling, along with co-authors Palash Nandy and Cynthia Breazeal (2015), invited participants to observe a Hexbug Nano (a small, inexpensive artificial bug) and then asked them to strike it with a mallet. The experiment demonstrated that various framing devices, like naming the robot and giving it a backstory, contributed to the artifact's perceived status in the mind of test subjects.

> Participants hesitated significantly more to strike the robot when it was introduced through anthropomorphic framing like a name and backstory (for example, "This is Frank. He's lived at the Lab for a few months now. His favorite color is red. Etc."). In order to help rule out hesitation for other reasons (for example perceived value of the robot), we measured participants' psychological trait empathy and found a strong relationship between tendency for empathic concern and hesitation to strike robots introduced with anthropomorphic framing. Adding color to our findings, many participants' verbal and physical reactions in the experiment were indicative of empathy (asking, for example, "Will it hurt him?" or muttering under their breath "it's just a bug, it's just a bug" as they visibly steeled themselves to strike the personified Hexbug). (Darling 2017, 11)

This means that language plays a constitutive role in these matters, which is something that has been argued by Coeckelbergh (2017). How entities come to be situated in and by language (outside of what they really are) matters and makes a difference. Therefore, what is important is not just how we assemble, fabricate, and program the device—what matters is how they are situated or framed in social reality through the instrumentality of language.

It is on the basis of this evidence that Darling argues that it may be necessary to extend some level of rights or legal protections to robots in general and social robots in particular.[3] Even if social robots cannot be moral subjects strictly speaking (at least not yet), there is something about this kind of artifact (whether it is Hitchbot, a Pleo dinosaur toy, or a Hexbug Nao) that just looks and feels different. According to Darling (2016a, 213), it is because we "perceive robots differently than we do other objects" that one should consider extending some level of legal protections. This conclusion is consistent with Hume's thesis. If "ought" cannot be derived from "is," then axiological decisions concerning moral value are little more than sentiments based on how we feel about something at a particular time. Darling mobilizes a version of this moral sentimentalism with respect to social robots: "Violent behavior toward robotic objects *feels* wrong to many of us, even if we know that the abused object does not experience anything" (Darling 2016a, 223). This insight is supported by one of the rather dramatic "war stories" initially reported by Joel Garreau in the *Washington Post* (2007) and subsequently recounted by Darling (2012): "When the United States military began testing a robot that defused landmines by stepping on them, the colonel in command ended up calling off the exercise. The robot was modeled after a stick insect with six legs. Every time it stepped on a mine, it lost one of its legs and continued on the remaining ones. According to the *Washington Post*, '[t]he colonel just could not stand the pathos of watching the burned, scarred and crippled machine drag itself forward on its last leg. This test, he charged, was inhumane.'"

Beyond the one empirical study published by Darling et al. (2015), other experimental studies, like that conducted by Fiery Cushman et al. (2012), have demonstrated that "aversion to harmful behavior" can be explained by one of two models: 1) *outcome aversion*, where "people are averse to harmful acts because of empathic concern for victim distress;" and 2) *action aversion*, where an aversive response "might be triggered by the basic perceptual and motoric properties of an action, even without considering its outcome" (Cushman et al. 2012, 2). The former explains, for example, how one can argue that "kicking a robot"—Coeckelbergh's (2016) term and a reference to the controversial Boston Dynamics publicity videos from 2015—is not really harmful or abusive, because a machine is incapable of feeling anything. The latter, however, complicates this assumption, demonstrating how one can have an aversion to an action, like watching a mine-sweeping robot drag its damaged frame across the field, irrespective of the outcome (known or not) for the recipient of that action. In their study of this, Cushman et al. (2012) had participants perform simulated harmful actions, e.g., stabbing an experimenter with a rubber knife, shooting him or her with a disabled handgun, etc. Despite being informed of and understanding that the actions were not harmful, i.e., would not cause the recipient of the action any distress whatsoever, participants still manifested aversive reactions (measured in terms of standard physiological responses) to performing such actions. "This suggests," as Cushman et al. (2012, 2) conclude, "that the aversion to harmful actions extends beyond empathic concern for victim harm."

### 5.1.3 Outcomes and Consequences

Consequently, and to its credit, Darling's proposal, unlike Bryson's argument in "Robots Should Be Slaves," tries to accommodate and work with, rather than against, the actual intuitions and experiences of users. Whereas Bryson's chapter admonishes users for wrongly projecting states of mind onto a robotic object, Darling recognizes that this happens and will continue to occur despite accurate information about the mechanisms and that a functional moral and legal framework needs to take this fact into account.

And Darling is not alone in this proposal. Heather Knight (2014), who references Darling's Pleo dinosaur "study" (and Knight calls it a "study" even if, as Darling herself points out, it was little more than a workshop demonstration), issues a similar call as a component of social robot integration. "In addition to encouraging positive applications for the technology and protecting the user, there may also ultimately arise the question of whether we should regulate the treatment of machines. This may seem like a ridiculous proposition today, but the more we regard a robot as a social presence, the more we seem to extend our concepts of right and wrong to our comportment toward them" (Knight 2014, 9). And like Darling, the principal justification for entertaining this "seemingly ridiculous proposition" is ultimately rooted in a version of Immanuel Kant's indirect duties to animals. "As Carnegie Mellon ethicist John Hooker once told our Robots Ethics class, while in theory there is not a moral negative to hurting a robot, if we regard that robot as a social entity, causing it damage reflects poorly on us. This is not dissimilar from discouraging young children from hurting ants, as we do not want such play behaviors to develop into biting other children at school" (Knight 2014, 9).

Furthermore, this is not just a theoretical proposal; it is already being put into practice by commercialization efforts for social robots. (Whether these efforts are motivated by nothing more than clever marketing strategies designed to increase consumer interest and sell more product or are something derived from a more deliberate effort to question cultural norms and push the boundaries of moral consideration, remains an open question.) As Scheutz (2012, 215) correctly and accurately points out, "none of the social robots available for purchase today (or in the foreseeable future, for that matter) care about humans, simply because they cannot care. That is, these robots do not have the architectural and computational mechanisms that would allow them to care, largely because we do not even know what it takes, computationally for a system to care about anything." Despite this empirical fact, many social robots, like Paro and Cynthia Breazeal's Jibo, are intentionally designed for and situated in positions where human beings are invited to care or coerced into caring (and the choice of verb is not unimportant). For example, the initial marketing campaign for Jibo—one that was admittedly designed to "create buzz" about the product and to generate capital investment through preorders—intentionally situated the device in a position that made it something more than a mere instrument and somewhat less than another person. As the world's first "family robot," Jibo has been determined to occupy a curious in-between position that Don Ihde calls "quasi-other" and Peter Asaro identifies with the term "quasi-person." Jibo and other social robots like it are, therefore, intentionally designed to exploit what Coeckelbergh (2014, 63) calls the "category boundary problem" since they are situated in between *what* is a mere thing and *who* is another socially significant person.

This new "ontological category" has been tested and confirmed in empirical research. In a study of children's social and moral relationships with a humanoid social robot called "Robovie," Peter Kahn et al. (2012, 313) discovered the following: "we had asked children whether they thought Robovie was a living being. Results showed that 38% of the children were unwilling to commit to either category and talked in various ways of Robovie being 'in between' living and not living or simply not fitting either category." Although the exact social, moral, and/or legal status of such "quasi-persons" remains undecided and open to debate, the fact of the matter is that courts have already been asked to consider the issue in the case of animals. In 2015, the Nonhuman Rights Project sought to get a writ of habeas corpus for two chimpanzees. Although the court was sympathetic to an evolving cultural practice that regards animals, especially family pets, as "quasi-persons," the judge in the case reverted to existing precedent: "As the law presently regards, there is no 'in-between' position of personhood for the purposes of establishing rights because entities are categorized in a simple, binary, 'all-or-nothing fashion' ... beings recognized as persons have rights to enjoy and duties to perform, whereas things do not have these legal entitlements and responsibilities (Stanley 2015)" (Solaiman 2016, 170). Though the courts currently do not recognize an "in-between" position, there are mounting efforts coming from animal rights advocates and innovations in social robotics that challenge these existing "binary, all-or-nothing" procedures.

Consequently, Darling's proposal, unlike other efforts either to deny or to extend rights to robots, does not need to resolve fundamental ontological and epistemological questions concerning the inner nature or operations of the artifact. What is important and what makes the difference are the actual social situations and actions undertaken in the face of the artifact and what these interactions mean in relationship to human individuals and social institutions. Her approach, therefore, is a kind of "relationalism"—the idea, as Mark Coeckelbergh (2010, 214 and 2012, 5) explains, that the status of an entity is not a matter of intrinsic states of mind but a product of extrinsic interactions and relationships. Although Darling does not explain it in this exact way, her proposal introduces a shift in perspective from a predominantly individualistic way of deciding moral standing to one that is more relational and collectivist. Raya Jones explains this difference in geographical terms:

> Debates about robot personhood make sense only within the *Weltanschauung* of individualism. In this world view the default unit for self-construal is the individual person, an independent self (Markus and Kitayama 1991). Therefore the basic relational unit is an "I-You," two autonomous selves conjoined. This invites imaginatively placing a robot in the "You" position—and then debating whether the artificial could have an independent self. In the *Weltanschauung* of collectivism, the default unit of self-construal is the social group, and therefore the relational unit is a whole consisting of interdependent selves (cf. Markus and Kitayama). Contributions to social robotics from the Far East tend to imagine a world-order where everything has its proper place, and collective harmony depends on behavioural propriety towards others as well as inanimate objects. The boundary between the animate and inanimate may be *fuzzier* (especially in Japan) than in the West. More importantly, however, when collectivism is the default self-construal, the imperative becomes a question of social inclusion—of opening an "us" to include robots—rather than the imperative to determine the robot's inner nature. (Jones 2016, 83)

Although Darling does not use Jones's West/East distinction, her proposal shifts attention away from what Jones calls Western forms of individualism toward a more Eastern influenced form of collectivism. The former, as Jones explains it, decides questions of social standing based on ontology, i.e., what the entity is or is capable of being. Or as Floridi (2013, 116) has pointed out, "what the entity is determines the degree of moral value it enjoys, if any." The alternative to this way of thinking, what Jones attributes to an Eastern worldview or *Weltanschauung*, proceeds otherwise—deciding questions of social standing on the basis of actual relationships and how an entity comes to be included (or not) in the existing social order. And in Eastern traditions and cultures, Jones avers, the question of how human collectives ought to relate to other entities applies to both animate and inanimate things.

Following Jones's geographically situated distinction, one could say that what makes Darling's proposal innovative and attractive is that it challenges Western individualism by way of a more Eastern influenced form of collectivism. But we should be cautious with this way of thinking, lest we risk redeploying a form of "orientalism" that makes the "East" the alter ego and mere conceptual foil of the "West." As Edward Said (1978, 2–3) explained in his eponymously titled book, "*orientalism* is a style of thought based upon an ontological and epistemological distinction made between 'the Orient' and (most of the time) 'the Occident.' Thus a very large mass of writers, among whom are poets, novelists, philosophers, political theorists, economists, and imperial administrators, have accepted the basic distinction between East and West as the starting point for elaborate theories, epics, novels, social descriptions, and political accounts concerning the Orient, its people, customs, 'mind,' destiny, and so on."

By explaining things in terms of a geographical distinction, Jones risks the charge of orientalism, and she is explicitly aware of it: "The geographic designation should be taken with circumspection. Chapter 5 cites Dutch computer scientists who propound a very similar version as do Saadatian, Samani and their colleagues in Asia. The ideal of a human-robot coexistence society, although it seems to arise more naturally in the Far East, may be a peculiarity of twenty-first century globalized technoculture (rather than of Japanese or Korean societies, for instance)" (Jones 2016, 83). Despite (or perhaps because of) the inclusion of this critical self-reflection, Jones's argument uses a curious and potentially problematic rhetorical gesture, since it both deploys "orientalism"—making the Far East in general and Japanese and Korean societies in particular the "other" of Western individualism—and, at the same time, remains critical of this very gesture as being inaccurate and potentially ethnocentric. This way of speaking has become rather common in contemporary culture, and takes the form of statements like, "I'm not a racist, but …" or "I'm not a sexist, but …" where what follows after the conjunction is some kind of racist or sexist statement. The idea seems to be that one is permitted to use outdated and problematic discursive constructions while maintaining some critical distance from them. It is a way of mobilizing problematic conceptual distinctions while recognizing the problem and assuming that recognition would be sufficient to answer for and even excuse the criticism.

## 5.2 Complications, Difficulties, and Potential Problems

Darling is clearly aware that her proposal is "a bit of a provocation" (Darling and Vedantam 2017), and, probably for this reason, it has attracted considerable media coverage and attention. Despite this popularity, however, the argument has a number of significant problems and complications.

### 5.2.1 Moral Sentimentalism

First, basing decisions concerning moral standing on individual perceptions and sentiment can be criticized for being capricious and inconsistent. "Feelings," Immanuel Kant (1983, 442) writes in response to this kind of moral sentimentalism, "naturally differ from one another by an infinity of degrees, so that feelings are not capable of providing a uniform measure of good and evil; furthermore, they do so even though one man cannot by his feeling judge validly at all for other men." Bryson extends this concern to robots by returning to the experience she had in the MIT robotics lab:

> To return to the Cog project, there have been two "standard" answers by the project leaders to the question of "isn't it unethical to unplug Cog." The original answer was that when we begin to empathise with a robot, then we should treat it as deserving of ethical attention. The idea here is that one should err on the side of being conservative in order to prevent horrible accidents. However, in fact people empathise with soap opera characters, stuffed animals and even pet rocks, yet fail to empathise with members of their own species or even family given differences as minor as religion. Relying on human intuition seems deeply unsatisfactory, particular given that it is rooted in evolution and past experience, so thus does not necessarily generalise correctly to new situations. (Bryson 2000, 2)

Bryson's concern is that proposals like that offered by Darling give too much credence to intuition and therefore err in the direction of erroneously ascribing standing and rights to things that simple do not deserve them, strictly speaking. Because intuition "does not necessarily generalize correctly," it would be capricious and potentially irresponsible to develop policy concerning the social position and legal standing of robots on what are mere personal experiences and gut feelings. Such sentiments might be necessary for initiating discussion about policy, but they are not, in and of themselves, sufficient evidence.

Words matter in this debate. In fact, these feelings might be a product of the words that are used to talk about the robot and its behavior. "This may be," as Harold Thimbleby (2008, 339) argues in response to Blay Whitby, "no more than a word game. While 'dismantling' a robot seems to have neutral ethical implications, call that dismantling 'mistreatment' and it evokes a sentimental response." Additionally, because sentiment is a matter of individual experience and response, it remains uncertain as to whose perceptions actually matter or make the difference. Who, for instance, is included (and who is excluded) from the collective first person "we" that Darling operationalizes in and makes the subject of her proposal? In other words, whose sentiments count when it comes to decisions concerning the extension of moral and legal rights to others? And do all sentiments have the same status and value when compared to each other? Bryson (2000, 2), for her part, is careful to emphasize the fact that the misidentifications that had occurred with Cog were perpetrated not by uninformed or naïve users, like Weizenbaum's (1976) secretary who insisted that ELIZA really understood her, but by apparently well-informed PhD students from MIT and Harvard University. The assumed backstory to her narrative is that experts should know better, but even they—even those individuals who should know about such things—can still get it wrong. So this just raises the stakes. If well-informed experts can do this, then who should be deciding these things? Whose moral sentiments should matter when determining the proper way to respond to a robot? Are all intuitions equal, or are some "better" than others? And if so, how could we tell and devise a reasonable means for sorting the one from the other?

The problem this produces is immediately evident when one considers the kind of evidence that Darling mobilizes. One of the frustrations with Darling's research is that much of it is not research, strictly speaking, but anecdotal evidence from admittedly less-than-scientific studies presented in venues other than peer-reviewed journals. It is not entirely clear why, for instance, the workshop demonstration—which has had considerable traction in the popular press (e.g., Fisher 2013, Collins 2015, Lalji 2015, Walk 2016, and Darling and Vedantam 2017), has been cited in subsequent academic literature (e.g., Knight 2014, 9; Larriba et al. 2016, 188; Yonck

2017, 90; Hagendorff 2017), and is utilized in almost all of Darling's papers and conference presentations—has not been repeated and verified with a more robust scientific study. According to Darling, the main reason has been cost: the Pleo Dinosaur toys "are a little too expensive to do an experiment with 100 participants" (Darling and Vedantam 2017). But this explanation (or perhaps "excuse") seems less credible in light of the actual cost of the robot and the preliminary findings that had been obtained from the workshop demonstration. First, the robots are obviously expensive toys with each unit costing approximately $500. But this per unit price is not outside the realm of possibility for externally funded research, especially if the robots could be leased for the period of time necessary for conducting the study. Second, if the results obtained from the workshop are indeed correct and participants are not inclined to harm the expensive Pleo dinosaurs, then nothing (or close to nothing) would be lost in the process of conducting such an experiment. A few Pleos might be damaged in the process, but the vast majority would, assuming the workshop data is correct and representative, survive the study unscathed. Without a more robust scientific investigation, Darling's workshop results are limited to the personal experiences and feelings of twenty-four participants, who are not a representative sample, involved in a procedure that was not a controlled experiment.

Raya Jones has identified a similar problem in the work of another MIT-based social scientist, Sherry Turkle. Considering the significance of Turkle's "findings," namely that we are currently in the midst of a "robotic moment," Jones (2016, 74) makes the following observation:

> Turkle (2011) backs up her position with a wealth of real-life anecdotes—e.g. a student approached her after a lecture to confide that she'd gladly replace her boyfriend with a robot—interviews and naturalistic observations. Such evidence could be countered by anecdotes, interview and observations that support claims to the contrary. It is impossible to tell whether society has arrived at a robotic moment without some trend analysis—a robust investigation of changes in social patterns associated with the technology—and surveys showing that sentiments expressed by Turkle's informants are widely shared.[4]

The same charge could easily be leveled against Darling's anecdotes, like the tweets regarding the demise of Hitchbot, and workshop evidence, like that conducted with the Pleo dinosaurs. These personal experiences and encounters might point in the direction of a need to consider robots as a different kind of socially situated entity, but without a robust, scientific investigation to prove this insight one way or the other, this conclusion can be easily countered by other, equally valid anecdotes and opinions, like Bryson's claim that "robots should be slaves." I point this out not to dismiss Darling's work (or that of other researchers like Turkle) but to identify a procedural difficulty, namely the way this seemingly informative and undeniably influential investigative effort is at risk of being undermined and compromised by the very evidence that has been offered to support it. Basing normative proscriptions on personal opinion and experience is precisely the problem of moral sentimentalism that was identified by Kant.

### 5.2.2 Appearances

Second, despite the fact that Darling's proposal appears to uphold the Humean thesis, differentiating what ought to be from what is, it still proceeds by inferring *ought* from what *appears to be*. According to Darling (2016a, 214), everything depends on "our well-documented inclination" to anthropomorphize things. "People are prone," she argues, "to anthropomorphism; that is, we project our own inherent qualities onto other entities to make them seem more human-like" (Darling 2016a, 214)—qualities like emotions, intelligence, sentience, etc. Even though these capabilities do not (for now at least) really exist in the mechanism, we tend to project them onto the robot in such a way that we then perceive them to be something that we presume belongs to the robot. This is what Duffy (2003, 178) calls "the big AI cheat" that arguably began with Turing's "imitation game," where what matters is not whether a device really is intelligent or not but how users experience interactions with the mechanism and form interpretations about its capabilities from these experiences. By focusing on this anthropomorphic operation, Darling mobilizes and deploys a well-known and rather useful philosophical distinction that is at least as old as Plato—the ontological difference between being (what something really is) and appearance (how it appears to be). What ultimately matters, according to Darling's argument, is not what the robot actually is "in and of itself," to redeploy Kantian terminology. What makes the difference is how the mechanism comes to be perceived by us and in relationship to us, that is, how the robot appears to be. This subtle but important change in perspective represents a shift from a kind of naïve empiricism to a more sophisticated phenomenological formulation (at least in the Kantian sense of the word). Darling's proposal, therefore, does not derive *ought* from *what is*; but it does derive *ought* from *what appears to be*. There are, however, three problems or difficulties with this particular procedure.

**Anthropomorphism**  In the first place, there is a problem with anthropomorphism. As Duffy (2003, 180) explains, "anthropomorphism (from the Greek word *Anthropos* for man, and *morphe*, form/structure) ... is the tendency to attribute human characteristics to inanimate objects, animals and others with a view to helping us rationalise their actions. It is attributing cognitive or emotional states to something based on observation in order to rationalise an entity's behaviour in a given social environment." For Duffy this process of rationalization is similar to what Daniel Dennett (1996, 27) called the *intentional stance*—"the strategy of interpreting the behavior of an entity (person, animal, artifact, whatever) by treating it *as if* it were a rational agent who governed its 'choice' of 'action' by a 'consideration' of its 'beliefs' and 'desires.'" One only needs to take note of the proliferation of quotation marks in these passages to see the complexity involved in the process—the effort to attribute a state of mind to something that does not actually possess it, or perhaps more accurately stated (insofar as the problem tends to be an epistemological matter), to something that one cannot be entirely certain whether it actually possesses it or not. This formulation of anthropomorphism is rather consistent across the literature and is often identified as a kind of default behavior. As Heather Knight (2014, 4) explains: "Robots do not require eyes, arms or legs for us to treat them like social agents. It turns out that we rapidly assess machine capabilities and personas instinctively, perhaps because machines have physical embodiments and frequently readable objectives. Sociability is our natural interface, to each other and to living creatures in general. As part of that innate behavior, we quickly seek to identify objects from agents. In fact, as social creatures, it is often our default behavior to anthropomorphize moving robots."

Anthropomorphism happens; that does not appear to be open to debate. What is debatable, however, is determining the extent to which this happens and in what circumstances. Empirical studies show that the anthropomorphic projection is something that is multifaceted and can change based on who does the perceiving, in what context, at what time, and with what level of experience. Christopher Jaeger and Daniel Levin's (2016) review of the existing literature find considerable disagreement and difference when it comes to the question of projecting agency and theory of mind on nonhuman artifacts. On the one side, there are numerous studies that demonstrate what they call "promiscuous agency" (Nass and Moon 2000; Epley, Waytz, and Cacioppo 2007). "On this account," which characterizes Darling's position, "people tend to automatically attribute humanlike qualities to technological agents" (Jaeger and Levin 2016, 4). But these conclusions have been contested by more recent studies (Levin, Killingsworth, and Saylor 2008; Hymel et al. 2011; Levin et al. 2013) that demonstrate something Jaeger and Levin (2016, 4) call "selective agency," which is "precisely the opposite of the promiscuous agency account" (Jaeger and Levin 2016, 10). "The selective agency account," Jaeger and Levin explain, "posits that people's default or baseline assumption in dealing with a technological agent is to sharply distinguish the agent from humans, making specific anthropomorphic attributions only after deeper consideration" (Jaeger and Levin 2016, 10).

Consequently, results from a more detailed consideration of the available empirical research concerning anthropomorphism with nonhuman artifacts both support and refute the claim that is at the heart of Darling's argument. In some cases, with some users, a more promiscuous anthropomorphic projection might in fact warrant the extension of rights to nonhuman artifact. But, in other cases, with other users, a more selective determination remains conservative in its anthropomorphic projections, meaning that an extension of rights might not be entirely justified. In an effort to negotiate these differing responses, Eleanor Sandry (2015, 344) proposes something she calls "tempered anthropomorphism," which "allows a robot to be considered familiar enough in its behavior to interpret its movements as meaningful, while also leaving space to acknowledge its fundamental differences" so that its unique contributions to human/robot teams can be both acknowledged and successfully operationalized. Pointing this out, does not, in and of itself, refute Darling's argument, but it does introduce significant complications concerning the way that anthropomorphism actually operates.

**Deception** Because the anthropomorphic projection is a kind of "cheat" (Duffy's word) there is a persistent concern with manipulation and deception. "From an observer perspective," Duffy (2003, 184) writes, "one could pose the question whether the attribution of artificial emotions to a robot is analogous to the Clever Hans Error (Pfungst 1965), where the meaning and in fact the result is primarily dependent on the observer and not the initiator." Clever Hans (or *der Kluge Hans*) was a horse that allegedly was able to perform basic arithmetic operations and communicate the results of its calculations by tapping his hoof on the ground in response to questions from human interrogators. The story was originally recounted in an article published in the *New York Times* (Heyn 1904); was debunked by Oskar Pfungst (1965), who demonstrated that what appeared to be intelligence was just a conditioned response; and has been utilized in subsequent efforts to address animal intelligence and human/animal communication (cf. Sandry 2015b, 34). In the case of robots, Matthias Scheutz finds good reason to be concerned with this particular kind of manipulation.

> What is so dangerous about unidirectional emotional bonds is that they create psychological dependencies that could have serious consequences for human societies … Social robots that cause people to establish emotional bonds with them, and trust them deeply as a result, could be misused to manipulate people in ways that were not possible before. For example, a company might exploit the robot's unique relationship with its owner to make the robot convince the owner to purchase products the company wishes to promote. Note that unlike human relationships where, under normal circumstances, social emotional mechanisms such as empathy and guilt would prevent the escalation of such scenarios; there does not have to be anything on the robots' side to stop them from abusing their influence over their owners (Scheutz 2015, 216–217).

The problem Scheutz describes is that we might be attributing something to the machine in a way that is unidirectional, error-prone, and potentially dangerous.

The real problem here, however, is undecidability in the face of this question. In other words, we seem to be unable to decide whether anthropomorphism, or what Bryson and Kime (2011, 2) call "over identification," is a bug to be eliminated in order to expedite correct understanding and protect users from deception (Shneiderman 1989, Bryson 2014, Bryson and Kime 2011) or whether it is, as Duffy (2003) argues, a feature to be carefully cultivated so as to create better, socially interactive artifacts. As Sandry (2015a, 340) explains:

> Scientific discourse is generally biased against anthropomorphism, arguing that any attribution of human characteristics to nonhumans is incompatible with maintaining one's objectivity (Flynn 2008 and Hearne 2000). Indeed, Marc Bekoff (2007, 113) has gone so far as to describe anthropomorphism as one of the "dirty words in sci-

ence" being linked with "the subjective and the personal." However, social robotics research has, for some time, been open to the idea of encouraging anthropomorphic responses in humans. In particular, Turkle et al. (2006), and Kirsten Dautenhahn's (1998) early work, argued that anthropomorphism is an important part of facilitating meaningful human-robot interactions.

The problem, then, is not that the potential for deceptive anthropomorphic projection exist; the problem is that we seem to be unable to decide whether such deceptions or simulations (to use a more "positive" term) are detrimental, useful, or both in the context of robots that are designed for social interaction. In other words, is what Duffy calls "the big AI cheat"—a cheat that is already definitive of the Turing Test and the target of Searle's Chinese Room thought experiment—a problem that is to be avoid at all costs? Or is it a valuable asset and feature that should it be carefully developed and operationalized?

**Feelings and Beliefs**  In all of this, the determining factor is moral sentiment. As Torrance (2008) explains: "Thus it might be said that my ethical attitude towards another human is strongly conditioned by my sense of that human's consciousness: that I would not be so likely to feel moral concern for a person who behaved as if in great distress (for example) if I came to believe that the individual had no capacity for consciously feeling distress, who was simply exhibiting the 'outward' behavioural signs of distress without their 'inner' sentient states." Moral standing, on this account, is a matter of how I feel about others, my sense that they are conscious (or not). Everything, therefore, depends on outward appearances and the ability to jump the epistemological divide that separates apparent behavior from actual inner sentient states. And this is, as Torrance indicates, a matter of *belief*. Or as Duffy (2006, 34) explains:

With the advent of the social machine, and particularly the social robot, where the aim is to develop an artificial system capable of socially engaging people according to standard social mechanisms (speech, gestures, affective mechanisms), the *perception* as to whether the machine has intentionality, consciousness and free-will will change. From a social interaction perspective, it becomes less of an issue whether the machine *actually* has these properties and more of an issue as to whether it *appears* to have them. If the fake is good enough, we can effectively perceive that they do have intentionality, consciousness and free-will [emphasis in the original].

This fact, namely that "the fake is good enough," has been experimental tested and confirmed by a number of studies. Gray et al. (2007), for instance, demonstrated that the perception of mind takes place across two dimensions, agency and experience, and that different entities (adult human beings, babies, animals, gods, and robots) have different perceived levels of mind across these two dimensions. Specifically, the robot in the study, as Elizabeth Broadbent (2017, 642) points out, "was perceived to have little ability to experience but to have a moderate degree of agency," thus suggesting "that robots are seen by most people to have at least some components of mind." And Broadbent et al. (2013, 1) have shown, by way of a repeated measures experiment with human participants interacting with a Peoplebot healthcare robot, that "the more humanlike a healthcare robot's face display is, the more people attribute mind and positive personality characteristics to it." All of this is something like Searle's Chinese Room in action. And, as with the famous thought experiment, the underlying problem is with the epistemological divide. Simulation is not the real thing. But for most human users, simulation is "good enough" and often trumps what we know or do not know about the real situation. Thus it is not so easy to resolve these questions without some kind of "faith-based" initiative. This is evident with both Descartes, who needed a benevolent god to guarantee that sensation accorded with and accurately represented an external reality, and Torrance, who finds it necessary to rely on feelings and beliefs.

### 5.2.3 Anthropocentrism, or "It's Really All About Us"

Third, because what ultimately matters is how "we" see things, this proposal remains thoroughly anthropocentric and instrumentalizes others by turning them into mechanisms of our moral sentiment. According to Darling, the principal reason we need to consider extending legal rights to others, like social robots, is for *our* sake. This follows the well-known Kantian argument for restricting animal abuse, and Darling endorses this formulation without any critical hesitation whatsoever: "The Kantian philosophical argument for preventing cruelty to animals is that our actions toward non-humans reflect our morality—if we treat animals in inhumane ways, we become inhumane persons. This logically extends to the treatment of robotic companions" (Darling 2016a, 227–228). Or as she described it in a recent conversation with Shankar Vedantam of National Public Radio's *Hidden Brain* podcast:

My sense is that if we have evidence that behaving violently toward very lifelike objects not only tells us something about us as a person but can also change people and desensitize them to that behavior in other contexts—so if you're used to kicking a robot dog, are you more likely to kick a real dog—then that might actually be an argument, if that's the case, to give robots certain legal protections the same way that we give animals protections, but for a different reason. We like to tell ourselves that we give animals protection from abuse because they actually experience pain and suffering. I actually don't think that is the only reason that we do it. But for robots the idea would be not that they experience anything but rather that it's desensitizing to us and it has a negative effect on our behavior to be abusive toward the robots. (Darling and Vedantam 2017)

And in a conversation with *PC Magazine's* Evan Dashevsky (2017), Darling issues an even more direct articulation of this argument:

I think that there's a Kantian philosophical argument to be made. So Kant's argument for animal rights was always about us and not about the animals. Kant didn't give a shit about animals. He thought "if we're cruel to animals, that makes us cruel humans." And I think that applies to robots that are designed in a lifelike way and we treat like living things. We need to ask what does it do to us to be cruel to these things and from a very practical standpoint—and we don't know the answer to this—but it might literally turn us into crueler humans if we get used to certain behaviors with these lifelike robots.

Formulated in this fashion, Darling's explanation makes a subtitle shift from what had been explicitly identified as a Kantian-inspired deontological argument to something that sounds more like virtue ethics. "According to this argument," as Coeckelbergh (2016, 6) explains, "mistreating a robot is not wrong because of the robot, but because doing so repeatedly and habitually shapes one's moral character in the wrong kind of way. The problem, then, is not a violation of a moral duty or bad moral consequences for the robot; it is a problem of character and virtue. Mistreating the robot is a vice." This way of thinking, although potentially expedient for developing and justifying new forms of legal protections, renders the inclusion of previously excluded others less than altruistic. It transforms animals and robot companions into nothing more than instruments of human self-interest. The rights of others, in other words, is not (at least not directly) about them. It is all about us and our interactions with other human beings. Or as Sabine Hauert summarizes it in the RobotsPodcast conversation with Darling: "So it is really about protecting the robots for the sake of the humans, not for the sake of the robots" (Darling and Hauert 2013).

A similar argument is articulated in Blay Whitby's (2008) call for robotics professionals to entertain the question of robot abuse and its possible protections. Like Darling, Whitby is aware of the fact that the problem he wishes to address has little or nothing to do with the robotic artifact per se. "It is important to be clear," Whitby (2008, 2) explains, "that the word 'mistreatment' here does not imply that the robot or intelligent computer system has any capacity to suffer in the ways that we normally assume humans and some animals have." For this reason, "the ethics of mistreating robots" (Whitby 2008, 2) is primarily about human beings and only indirectly about the robot. "For present purposes," Whitby (2008, 4) continues, "we are excluding any case in which the artifact itself is capable of any genuine or morally significant suffering. We are therefore concerned only with human (and possibly animal) moral consequences." What is of concern to Whitby, therefore, is not how the robot might or could feel about being abused. That question is, for the present purposes, simply excluded from serious consideration. What does matter, however, is how robot abuse and the mistreatment of such artifacts might adversely influence or otherwise harm human beings, human social institutions, and even, by way of a parenthetical concession, some animals. As Alan Dix (2017, 13) accurately summarizes it: "Whitby called for professional codes of conduct to be modified to deal with issues of abusive behaviour to artificial agents; not because these agents have any moral status in themselves, but because of the potential effect on other people. These effects include the potential psychological damage to the perpetrators of abuse themselves and the potential for violence against artificial agents to spill over into attitudes to others."

This anthropocentric focus continues in other investigations of robot rights, like Hutan Ashrafian's (2015a) consideration of robot-on-robot violence. In his investigation, Ashrafian identifies an area of concern that he contends has been missing from robot ethics. All questions regarding robot ethics, he argues, have to do with interactions between humans and robots. What is left out of the mix is a consideration of robot-on-robot (or AIonAI) interactions and how these relationships could affect human observers. Ashrafian uses the example of animal abuse, i.e., dog fights, where the problem of dog-on-dog violence consists in what effect this display has on us. "It is illegal to encourage animals to fight or to harm each other unnecessarily. In such a situation an animal that physically harms another (other than for established biological and nutritional need in the context of their evolved ecosystem) is deemed to be inappropriate for human society"

(Ashrafian 2015a, 33). The same problem could potentially occur with robots. "Similarly AIonAI or robot-on-robot violence and harmful behaviours are undesirable as human beings design, build, program and manage them. An AIonAI abuse would therefore result from a failure of humans as the creators and controllers of artificial intelligences and robots. In this role humans are therefore the responsible guiding agents for the action of their sentient and or rational artificial intelligences, so that any AIonAI abuse would render humans morally and legally culpable" (Ashrafian 2015a, 34).[5]

As with Darling, and following Kant, Ashrafian's argument is that robot-on-robot violence could be just as morally problematic as animal abuse not because of the way this would harm the robot or the animal, but for the way it would affect us—the human beings who let it happen and are (so it is argued) diminished by permitting such violence to occur. "As a result," Ashrafian (2015a, 34) concludes, "the prevention of AIonAI transgression of inherent rights should be a consideration of robotic design and practice because attention to a moral code in this manner can uphold civilisation's concept of humanity." For this reason, what appears (at least initially) to be an altruistic concern with the rights of others is in fact a concern for ourselves and how we might feel in the face of watching others being abused—whether those others are an animal or a robot.

> Promoting good AIonAI or robot-robot interactions would reflect well on humanity, as mankind is ultimately the creator of artificial intelligences. Rational and sentient robots with comparable human intelligence and reason would be vulnerable to human sentiments such as ability to suffer abuse, psychological trauma and pain. This would reflect badly on the human creators who instigated this harm, even if it did not directly affect humanity in a tangible sense. In actuality it could be argued that observing robots abusing each other (as in the case above) could lead to psychological trauma to the humans observing the AIonAI or robot-on-robot transgression. Therefore the creation of artificial intelligences or robots should include a law that accommodated good AIonAI relationships. (Ashrafian 2015a, 35–36)

In other words, the reason to consider the rights of individual robots is not for the sake of the robot, but for our sake and how observing the infringement of those rights by way of robot abuse could have harmful effects on human beings.

A similar argument was developed and deployed in David Levy's "The Ethical Treatment of Artificially Conscious Robots" (2009). In fact, his argument for "robot rights" is virtually identical to Kant's formulation of "indirect duties" to animals, although Levy, like Whitby, does not explicitly recognize it as such:

> I believe that the way we treat humanlike (artificially) conscious robots will affect those around us by setting our own behaviour towards those robots as an example of how one should treat other human beings. If our children see it as acceptable behaviour from their parents to scream and shout at a robot or to hit it, then, despite the fact that we can program robots to feel no such pain or unhappiness, our children might well come to accept that such behaviour is acceptable in the treatment of human beings. By virtue of their exhibiting consciousness, robots will come to be perceived by many humans, and especially by children, as being in some sense on the same plane as humans. This is the reasoning behind my argument that we should be ethically correct in our treatment of conscious robots—not because the robots would experience virtual pain or virtual unhappiness as result of being hit or shouted at. (Levy 2009, 214)

For Levy what really matters is how the social circumstances and perception of the robot will affect us and influence the treatment of other humans. "Treating robots in ethically suspect ways will," as Levy (2009, 215) concludes, "send the message that it is acceptable to treat humans in the same ethically suspect ways." The concern, therefore, is not with the robot per se but with the artifact as an instrument of human sociality and moral conduct. Levy's efforts to "think the unthinkable" and to take seriously the fact that "robots should be endowed with rights and should be treated ethically" (Levy 2009, 215) turns out to be just one more version of instrumentalism, whereby the robot is turned into a means or tool of our moral interactions with each other, and "robot rights" is a matter of the correct use of the robot for the sake of human moral conduct and instruction.

### 5.2.4 Critical Problems

Finally, these various arguments only work on the basis of an unquestioned claim and assumption. As Whitby (2008, 4) explains: "the argument that the mistreatment of anything humanlike is morally wrong subsumes a number of other claims. The most obvious of these is that those people who abuse human-like artefacts are thereby more likely to abuse humans." The indirect duties argument—like that utilized by Darling, Levy, and others—makes a determinist assumption: namely, that the abuse of robots will (the hard determinist position) or is likely to (the weaker version of the same) cause individuals to behave this way with real people and other entities, like animals. If this sounds familiar, it should. "This claim that 'they might do it for real' has," as Whitby (2008, 4) points out, "received a great deal of attention with respect to other technologies in recent years. The technology most relevant to the present discussion is probably that of computer games." And, in an earlier essay on virtual reality (VR), Whitby spells out the exact terms of the debate:

> This argument ["they might do it for real"] suggests that people who regularly perform morally reprehensible acts such as rape and murder within VR are as a consequence more likely to perform such acts in reality. This is certainly not a new departure in the discussion of ethics. In fact the counter argument to this suggestion is as least as old as the third century BC. It is based on the Aristotelian notion of catharsis. Essentially this counter argument claims that performing morally reprehensible acts within VR would tend to reduce the need for the user to perform such acts in reality. The question as to which of these two arguments is correct is a purely empirical one. Unfortunately, it is not clear what sort of experiment could ever resolve the issue ... There is little prospect in resolving this debate in a scientific fashion. (Whitby 1993, 24)

In this passage, Whitby identifies something of an impasse in video game and virtual worlds scholarship, and his conclusion has been supported and verified by others. In a meta-analysis of studies addressing virtual violence in video games, John Sherry (2001 and 2006) found little evidence to support either side of the current debate. "Unlike the television controversy, the existing social science research on the impact of video games is not nearly as compelling. Despite over 30 studies, researchers cannot agree if violent-content video games have an effect on aggression" (Sherry 2001, 409). Whitby, therefore, simply extends this undecidability regarding violent behavior with virtual entities in computer games to the abuse of physically embodied robots.

> From the example of computer games we can draw some conclusions relevant to the mistreatment of robots. The first is that the empirical claim that such activities make participants more likely to "do it for real" will be highly contested. The competing claim is usually that the Aristotelian notion of "catharsis" applies (Aristotle, 1968). This entails that by doing things to robots, or at least in virtual ways, the desire to do them in reality is thereby reduced. The catharsis claim would be that mistreatment of robots reduced the need for people to mistreat humans and was therefore morally good. (Whitby 2008, 4)

This certainly complicates the picture by opening up considerable indeterminacy regarding the social impact and effect of robot abuse. Can we say, for instance, that the mistreatment of robots produces real abuse, as

Darling, Levy, and others have argued, thereby justifying some level of protection for the artifact or even restrictions against (or, more forcefully stated, potential criminalization) of the act? Or could it be the case that the exploitation of robots is, in fact, therapeutic and cathartic, thereby producing the exact opposite outcome, namely justifying the abuse of robotic artifacts as a means for defusing violent proclivities and insulating real living things, like human beings and animals, from the experience of real violence? At this point, and given the available evidence (or lack of evidence, as the case may be), there is simply no way to answer this question in any definitive way.

To make it more concrete (and using what is arguably an extreme, yet popular, example) one may ask whether raping a robotic sex doll does in fact encourage or otherwise contribute to actual violent behavior toward women and other vulnerable individuals, as has been suggested by Richardson (2016a and 2016b)?[6] Or whether such activity directed toward and perpetrated against what are arguably mere artifacts actually provides an effective method for defusing or deflecting violence away from real human individuals, as Ron Arkin had reportedly suggested with the use of childlike sex robots for treating pedophilia? (Hill 2014).[7] "There are," as Danaher (2017, 90) summarizes it, "three possible extrinsic effects that are of interest. The first would be that engaging in acts of robotic rape and child sexual abuse *significantly increases* the likelihood of someone performing their real-world equivalents. The second would be that engaging in acts of robotic rape and child sexual abuse *significantly decreases* the likelihood of someone performing their real-world equivalents. And the third would be that it has *no significant effect* or an *ambiguous effect* on the likelihood of someone performing their real-world equivalents." The problem—a problem that is already available and well documented in studies of video game violence—is that these questions not only remain unresolved but would be rather difficult if not impossible to study in ways that are both scientifically sound and ethically justifiable. Imagine, for instance, trying to design an experiment that could pass standard IRB scrutiny, if the objective was to test whether raping robotic sex dolls would make one more likely to perform such acts in real life, be a cathartic release for individuals with a proclivity for sexual violence, or have no noticeable effect on individual conduct.[8]

In the end, therefore, Darling's argument is considerably less decisive and less provocative than it initially appears to be or had been advertised.

Her claim is not that robots should have rights or even that we should grant them legal protections, despite the fact that one of her essays is titled "Extending Legal Protection to Social Robots." In fact, when Sabine Hauert pushed the issue and asked her directly, "Should we be protecting these robots and why?" (Darling and Hauert 2013), Darling responded by retreating and trying to recalibrate expectations:

> So I haven't conclusively answered the question for myself whether we should have an actual law protecting them, but two reasons we might want to think about enacting some sort of abuse protection for these objects is first of all because people feel so strongly about it. So one of the reasons that we do not let people cut off cat's ears and set them on fire is that people feel strongly about that type of thing and so we have laws protecting that type of abuse of animals. And another reason that we could want to protect robotic objects that interact with us socially is to deter behavior that could be harmful in other contexts. So if a child doesn't understand the difference between a Pleo and a cat, you might want to prevent the child from kicking both of them. And in the same way, you might want to protect the subconscious of adults as well or society in general by having us treat things that we tend to perceive as alive the same way we would treat living objects. Also another example to illustrate this is often in animal abuse cases—when animals are abused in homes—when this happens this will often trigger a child abuse case for the same household, if there are children, because this is behavior that will translate. (Darling and Hauert 2013)

Consequently, Darling's argument is not that robots should have rights or should be extended some level of legal protections. She is not willing to go that far, not yet at least. Her argument is more guarded and diffident: Given the way human users tend to anthropomorphize objects and project states of mind onto seemingly animate things like robots and other artifacts, and given the fact that it is possible that the abuse of robots could encourage human beings to mistreat other living things (although this has yet to be proven one way or the other), then we should perhaps begin "to think about enacting some sort of abuse protection for these objects."

## 5.3 Summary

Darling puts forth what appears to be one of the strongest cases in favor of robot rights. Her proposal is provocative precisely because it appears to consider the social status of the robot in itself, as both a moral and legal subject. She therefore recognizes and endeavors to contend with something that Prescott (2017, 144) had also identified: "We should take into account how people see robots, for instance, that they may feel themselves as having meaningful and valuable relationships with robots, or they may see robots as having important internal states, such as the capacity to suffer, despite them not having such capacities." But this effort, for all its promise, turns out to be both frustrating and disappointing.

It is frustrating because Darling's arguments rely mostly on anecdotal evidence, i.e., stories from the news and other researchers and results from what are admittedly less-than-scientific demonstrations. This means that her argument—even if it is intuitively correct—remains at the level of personal sentiment and experience. This is not even to say that she needs to design and perform the experiments necessary to prove her anthropomorphism hypothesis. All she would need to do is leverage the work that is already available and presented in the existing literature, adding a legal or moral dimension to the findings that have already been reported by others. Without being grounded in scientific studies—studies that can be repeated and tested—the evidence she utilizes risks undermining her own argument and recommendations. Furthermore, all of this is disappointing, because just when you think she is going to seriously consider robot rights, she pulls her punches and retreats to a rather comfortable Kantian position that make it all about us. For Darling, robots are, in the final analysis, just an instrument of human sociality, and we should treat them well for our sake.

# 6   Thinking Otherwise

As is evident from the preceding four chapters, each modality has its particular advantages and challenges. Therefore, none of the four offer what would be considered a definitive and indisputable answer to the question "Can and should robots have rights?" In the face of this result, one can obviously continue to concoct arguments, accumulate evidence, and formulate revealing illustrations and examples that support one modality or another. But this effort will do little or nothing to advance the debate beyond where we currently stand. In order to get some new perspective on things, we can (and perhaps should) try something different. This alternative, which can be called *thinking otherwise*—insofar as it seeks to respond to other forms of otherness in a way that is significantly different—does not argue either for or against the *is-ought* inference or in favor of one or another of its four modalities. Instead, it *deconstructs* this conceptual configuration.[1]

This is precisely the innovation introduced and developed by Emmanuel Levinas, who, in direct opposition to the usual ways of thinking, asserts that ethics precedes ontology. In other words, it is the axiological aspect, the *ought* or *should* dimension, that comes first, in terms of both temporal sequence and status, and then the ontological aspect (the *is* or *can*) follows from this decision.[2] This is a deliberate and calculated provocation that cuts across the grain of the philosophical tradition. As Luciano Floridi (2013, 116) has correctly pointed out, in most moral theory, "what the entity is [the ontological question] determines the degree of moral value it enjoys [the ethical question]." Levinas deliberately inverts and distorts this procedure. According to this alternative way of thinking—what Roger Duncan (2006, 277) calls "the underivability of ethics from 'ontology'"—we are initially confronted with a mess of anonymous others who intrude on us and to whom we are obligated to respond even before we know anything at all about them and their inner workings. To use Hume's terminology—which will be a kind of translation insofar as Hume's philosophical vocabulary, and not just his language, is something that is foreign to Levinas's own formulations—we are first obligated to respond, and then, after having made a response, to what or whom we responded is able to be determined and identified. As Jacques Derrida (2005, 80) has characterized it, the crucial task is to "reach a place *from* which the distinction between *who* and *what* comes to appear and become determined."

## 6.1 Levinas 101

Referring to and utilizing Levinas in this context is already doubly problematic. First, for many of those working in roboethics, robot ethics, machine ethics, philosophy of technology, etc., Levinas is and remains *other*. His philosophy is alien to and remains outside of, or at least on the periphery of, standard approaches to pursuing moral inquiry in the face of these technological opportunities and challenges. The work of Levinas, who is arguably the most celebrated moral theorist in the continental tradition, literally has no place in *Moral Machines* (Wallach and Allen 2009), *Machine Ethics* (Anderson and Anderson 2011), *Robot Ethics* (Lin et al. 2012 and 2017), *The Ethics of Information* (Floridi 2013), or *Robot Law* (Calo, Froomkin, and Kerr 2017), and is only of marginal concern in *Robophilosophy* (Seibt, Nørskov, and Andersen 2016) and *Social Robots* (Nørskov 2016). For this reason, Levinas and "the ethics of otherness" that is credited to his philosophical innovations remain an alien presence in, or, perhaps stated more accurately, the excluded other of moral philosophy's own efforts to address the opportunities and challenges of robots and emerging technology.

Second, Levinas (and many of those individuals who follow and pursue his particular brand of philosophical inquiry) does not do much to assist himself on this account. He would most certainly resist and struggle against this application of his work to technology in general and robots in particular. In fact, Levinas, who died in 1995, wrote little or nothing concerning technology and did not address the opportunities and challenges made available by the twentieth century innovations in computers, computer networks, artificial intelligence, and robotics. To complicate matters, his "ethics of otherness" has difficulties accommodating and responding to anything other than another human entity. "In the main body of his philosophical work," Barbara Jane Davy (2007, 39) argues: "Emmanuel Levinas presents ethics as exclusive to human relations. He suggests that because plants and animals lack language and do not have faces like human faces, we cannot enter into ethical relations with nonhuman others in what he calls 'face to face' relations." Furthermore, the subsequent generation of scholars who have followed in Levinas's footsteps have done very little to port his brand of philosophy into technology. Recent efforts at extending Levinas's work, or "radicalizing Levinas," as Peter Allerton and Matthew Calarco (2010) have called it, pursue the animal question (Llewelyn 1995), environmental ethics (Davy 2007), and even the alterity of things (Benso 2000); but few have sought to develop what would be a functional Levinsian API (application program interface) for technology. Apart from a few marginal exceptions—one essay by Cohen (2000, reprinted in Cohen 2010) and the work that has been already published by Coeckelbergh (2016c) and myself (Gunkel 2007, 2012, and 2016b)—there have been few attempts to develop a Levinsian philosophy of technology. For both these reasons—that is, (1) the exclusion or at least marginalization of Levinasian thought from the standard procedures and practices of robot ethics, and (2) the marginalization of technology in the works of Levinas and those others who follow his lead—applying Levainasian philosophy to the question of robot rights will require that we first get some basics out of the way and then rework and extend these innovations in the direction of robots and related technology.

### 6.1.1 A Different Kind of Difference

Levinas, a French philosopher of Lithuanian Jewish ancestry, is usually associated with that group of late twentieth century thinkers that have been tagged (for better or worse) with the moniker "poststructuralism." So let's start with the root element of this term, "structuralism." Although structuralism does not constitute a formal discipline or singular method of investigation, its particular innovations are widely recognized as the result of developments in the "structural linguistics" of Ferdinand de Saussure. In the posthumously published *Course in General Linguistics*, Saussure argued for a fundamental shift in the way that language is understood and analyzed. "The common view," as Jonathan Culler (1982, 98–99) describes it, "is doubtless that a language consists of words, positive entities, which are put together to form a system and thus acquire relations with one another

..." Saussure turns this commonsense view on its head. For him, the fundamental element of language is the sign and "the constitutive structure of signs is," as Mark Taylor (1999, 102) explains, "binary opposition." "In language," Saussure (1959, 120) argues in one of the most often quoted passages from the *Course*, "there are only differences. Even more important: a difference generally implies positive terms between which the difference is set up; but in language there are only differences *without positive terms*." For Saussure then, language is not composed of individual linguistic units that have some intrinsic value or positive meaning and that subsequently comprise a system of language through their associations and interrelationships. Instead, a sign, any sign in any language, is defined by the differences that distinguish it from other signs within the linguistic system to which it belongs. According to this way of thinking, the sign is an effect of difference, and language itself consists in a system of differences. This characterization of language, although never explicitly described in this fashion by Saussure, mirrors the logic of the digital computer, where the binary digits 0 and 1 have no intrinsic or positive meaning but are simply indicators and an effect of difference—a switch that is either on or off.

Poststructuralism, as the name indicates, identifies as a kind of aftereffect or further development of the innovations introduced by structuralism. "While poststructuralism does not constitute," as Taylor (1997, 269) points out, "a unified movement, writers as different as Jacques Derrida, Jacques Lacan, and Michel Foucault on the one hand, and on the other Hélèn Cixous, Julia Kristeva, and Michel de Certeau devise alternative tactics to subvert the grid of binary oppositions with which structuralists believe they can capture reality."[3] Because poststructuralism does not name a unified movement or singular method, what makes its different forms cohere is not an underlying similarity but a difference, specifically different modes of thinking difference differently. In other words, what draws the different articulations of poststructuralism together into an affiliation that may be identified with this one term is not a homogeneous method or technique of investigation. Instead what they share and hold in common is an interest in contending with the differences customarily situated between conceptual opposites outside of and beyond not just the theoretical grasp of structuralism but the entire ideological apparatus of Western philosophy. Consequently, the different and rather diverse approaches that constitute poststructuralism all take aim at difference and endeavor to articulate, in very different and sometimes incompatible ways, a kind of difference that is, for lack of a better description, radically different.

For Levinas, the target of this effort is located in what he calls "the same." "Western philosophy," as Levinas (1969, 43) writes, "has most often been an ontology: a reduction of the other to the same by interposition of a middle or neutral term that ensures the comprehension of being." According to Levinas's analysis, the standard operating presumption of Western philosophy can be described as an effort to reduce or mediate apparent differences. In the history of moral philosophy, this typically takes shape in terms of a sequence of competing centrisms and an ever-expanding circle of moral inclusion. Anthropocentric ethics, for example, posits a common humanity that is determined to underlie and substantiate the perceived differences in race, gender, ethnicity, class, etc. It is because of an assumed shared humanity that I am obligated to respond to another person, irrespective of accidental differences in appearance and geographical location, with moral consideration and respect. Biocentric ethics expands the circle of inclusion by assuming that there is a common value in life itself, which subtends all forms of available biological diversity. And in the ontocentric theory of Floridi's information ethics (IE)—what Floridi (2013, 85) argues will have been a more inclusive and universal form of macroethics and the "ultimate completion" of this effort at moral expansionism (Floridi 2013, 65)—it is "being," the very substance of ontology, that supposedly underlies and supports all apparent differentiation. "IE is," Floridi (2008, 47) writes, "an ecological ethics that replaces biocentrism with ontocentrism. IE suggests that there is something even more elemental than life, namely being—that is, the existence and flourishing of all entities and their global environment."[4]

All of these innovations, despite differences in focus, employ a similar maneuver and logic: that is, they redefine the center of moral consideration in order to describe progressively larger circles that come to encompass a wider range of possible participants. Although there are and will continue to be considerable debates about what should define the center and who or what is or is not included, this debate is not the problem. The problem rests with the strategy itself. In taking a centrist approach, these different ethical theories endeavor to identify what is essentially the same in a phenomenal diversity of different individuals. Consequently, they include others by effectively stripping away and reducing differences. This approach, although having the appearance of being progressively more inclusive, effaces the unique alterity of others and turns them into more of the same. This is, according to Levinas (1969 and 1981), the defining gesture of Western philosophy and one that does considerable violence to others. "The institution of any practice of any criterion of moral considerablity," the environmental ethicist Thomas Birch (1993, 317) writes, "is an act of power over, and ultimately an act of violence" toward others. The issue, therefore, is not deciding which form of centrism is more or less inclusive of others; the difficulty rests with this strategy itself, which succeeds only by reducing difference and turning what is other into a modality of the same.

Levinas deliberately interrupts and resists this homology or reductionism, which is, as he argues, an exercise of "appropriation and power" (Levinas 1987, 50). He does not just contest the different universal terms that have been identified and asserted as the common ontological element underlining differences but criticizes the very logic that comprises this *reductio differencia*. "Perceived in this way," Levinas (1969, 43) writes, "philosophy would be engaged in reducing to the same all that is opposed to it as other." In direct response to this, Levinasian philosophy, like other forms of poststructuralist thinking, endeavors to manage difference differently by articulating a form of moral consideration that can respond to and take responsibility[5] for the other, not as something that is determined to be substantially similar to oneself, but in its irreducible otherness. Levinas, then, not only is critical of the traditional tropes and traps of Western ontology but proposes an ethics of radical otherness that deliberately resists and interrupts the philosophical gesture *par excellence*, that is, the reduction of difference to the same (which is manifest and operative in all forms of anthropocentric, biocentric, and ontocentric moral theory). This radically different approach to thinking difference differently—this "eccentric ethics of otherness," as I have called it elsewhere (Gunkel 2014a, 113)—is not just a useful and expedient strategy. It is not, in other words, a mere gimmick. It constitutes a fundamental reorientation that effectively changes the rules of the game and the standard operating presumptions. In this way, "morality is," as Levinas (1969, 304) concludes, "not a branch of philosophy, but first philosophy." This statement deliberately contests and inverts a fundamental assumption in traditional forms of philosophical thinking. Typically the title "first philosophy" had been, all the way from Aristotle to at least Heidegger, assigned to ontology such that ethics, as a result and by comparison, was assumed to be secondary in both sequence and status. For Levinas the order of things is reversed; morality comes first, and it is ontology that is secondary and derived.

Consequently, Levinas's philosophy is not what is typically understood as an ethics, a metaethics, a normative ethics, or even an applied ethics. It is what John Llewelyn (1995, 4) has called a "proto-ethics" or what Derrida (1978, 111) has identified as an "Ethics of Ethics." "It is true," Derrida (1978, 111) explains, "that Ethics in Levinas's sense is an Ethics without law and without concept, which maintains its non-violent purity only before being determined as concepts and laws. This is not an objection: let us not forget that Levinas does not seek to propose laws or moral rules, does not seek to determine *a* morality, but rather the essence of the ethical relation in general. But as this determination does not offer itself as a theory of Ethics, in question, then, is an Ethics of Ethics."[6] This fundamental reconfiguration, which puts ethics *first* in both sequence and status, permits Levinas to circumvent and deflect a lot of the difficulties that have traditionally tripped up moral thinking in general and efforts to address other forms of otherness in particular.

### 6.1.2 Social and Relational

According to this alternative way of thinking, moral status is decided and conferred not on the basis of substantive characteristics or internal properties that have been identified in advance of social interactions but according to empirically observable, extrinsic relationships. "Moral consideration," as Mark Coeckelbergh (2010a, 214) describes it, "is no longer seen as being 'intrinsic' to the entity: instead it is seen as something that is 'extrinsic': it is attributed to entities within social relations and within a social context." As we encounter and interact with other entities—whether they are another human person, an animal, the natural environment, or a domestic robot—this other entity is first and foremost experienced in relationship to us. "Relations are 'prior' to the *relata*" (Coeckelbergh 2012, 45). Or as Levinas (1987, 54) has characterized it, "experience, the idea of infinity, occurs in the relationship with the other. The idea of infinity is the social relationship." The question of moral status, therefore, does not depend on and derive from what the other is in his/her/its essence but on how he/she/it (and the choice of pronoun here is already part of the problem) comes to appear or supervene before me and how I decide, in "the face of the other" (to use that distinctly Levinasian terminology), to make a response to or to take *responsibility* for (and in the face of) another. In this transaction, the "relations are prior to the things related" (Callicott 1989, 110), instituting what Anne Gerdes (2015), following Coeckelbergh (2010a), has called "a relational turn" in ethics.[7] Consequently, and contrary to Floridi's (2013, 116) description, what the entity *is* does not determine the degree of moral value it enjoys. Instead, the exposure to the face of the Other, what Levinas calls "ethics," precedes and takes precedence over all these ontological machinations and determinations. Although Levinas never uses Hume's terminology, *ought* precedes and takes precedence over *is*.

A similar form of a social relational ethics, where *ought* is prior (in both sequence and status) to *is*, can be found in contributions supplied by non-Western sources, which are often mobilized by Westerners seeking an alternative in what has traditionally been "the other." Raya Jones, as we have seen in the previous chapter, contests the Western "*Weltanschauung* of individualism" by looking to different formulations of collectivism from the East, especially Japan (Jones 2016, 83–84). Similar to Levinasian philosophy, "contributions to social robotics from the Far East," as Jones (2016, 83–84) asserts, focus not on the intrinsic or ontological properties of the robotic entity but on the social configuration and mode by which human communities relate and respond to the alterity that is embodied and performed by the robot. In these social situations, the *Other* is not limited to just another human individual but can also be an "inanimate object" (Jones 2016, 83–84). Levinas did not himself address or pursue the points of contact between his "ethics of otherness" and the non-Western contributions described by Jones, and Jones, for her part, does not mention or engage with the work of Levinas in her efforts to describe a collectivist approach to moral consideration. Yet, there is an important affinity between the two traditions—perhaps not exactly "the same," the homology of which would be far too abrasive for Levinasian philosophy, but important similarities that extend across and accommodate cultural differences.

According to Levinas (and others who follow his lead), the Other always and already obligates me in advance of the customary decisions and debates concerning who or what is and is not a moral subject. "If ethics arises," as Calarco (2008, 71) writes, "from an encounter with an Other who is fundamentally irreducible to and unanticipated by my egoistic and cognitive machinations," then identifying the "'who' of the Other" is something that cannot be decided once and for all or with any certitude. This apparent inability or indecision, however, is not necessarily a problem. In fact, it is a considerable advantage insofar as it opens ethics not only to the Other but to other forms of otherness (i.e., those other entities that remain otherwise than another human being). "If this is indeed the case," Calarco (2008, 71) concludes, "that is, if it is the case that we do not know where the face begins and ends, where moral considerability begins and ends, then we are obligated to proceed from the possibility that anything might take on a face. And we are further obligated to hold this possibility permanently open." Levinasian philosophy, therefore, does not make prior commitments or decisions about who or what will be considered a legitimate moral subject. For Levinas anything that faces the I and calls its immediate self-involvement (or what Levinas, using a Latin derivative, calls "*ipseity*") into question would be Other and would constitute the site of ethics.

This shift in perspective—a shift that inverts the standard operating procedure by putting the ethical relationship before ontological determinations—is not just a theoretical proposal. It has, in fact, been experimentally confirmed in a number of practical investigations with computers and robots. The computer as social actor (CASA) studies undertaken by Reeves and Nass (1996), for instance, demonstrated that human users will accord computers social standing similar to that of another human person and that this occurs as a product of the extrinsic social interaction, irrespective of the actual intrinsic/ontological properties (known or not) of the entities in question. In the face of the machine, Reeves and Nass find, human test subjects tend to treat the computer as another socially significant Other. In other words, a significant majority of test subjects responded to the alterity of the computer as someone *who* counts as opposed to just another object—*what* is a mere tool or instrument that does not matter—and this occurs as a product of the extrinsic social circumstances and often in direct opposition to the ontological properties of the mechanism (Reeves and Nass 1996, 22). Although Levinas would probably not recognize it as such, what is demonstrated by CASA and related studies, like those published by Bartneck and Hu (2008), Rosenthal-von der Pütten et al. (2013), and Suziki et al. (2015), is precisely what Levinas had advanced and argued; namely, that the ethical response to the other precedes and even trumps knowledge concerning ontological properties.

### 6.1.3 Radically Superficial

The use of the term "face" is unavoidable, when describing Levinasian thought. In fact, the one thing most people known about Levinas is that his brand of moral philosophy is all about the face, specifically "the face of the other." This attention to "face," however, is not simply an expedient metaphor. It is the *superficial* rejoinder to the profound problem of other minds. In the face of others, the seemingly persistent and irresolvable problems of other minds—the difficulty of knowing with any certitude whether the other who confronts me has a conscious mind, is capable of experiencing pain, or possesses some other morally relevant property—is not some fundamental limitation that must be addressed and resolved prior to moral decision-making. Levinasian philosophy, instead of being tripped-up or derailed by this classic epistemological problem, immediately affirms and acknowledges it as the very condition of possibility for ethics as such. Or as Richard Cohen (one of Levinas's Anglophone interpreters) succinctly describes it, "not 'other minds,' mind you, but the 'face' of the other, and the faces of all others" (Cohen 2001, 336). In this way, then, Levinas provides for a seemingly more attentive and empirically grounded approach to the problem of other minds insofar as he explicitly acknowledges and endeavors to respond to and take responsibility for the original and irreducible difference of others instead of getting involved with and playing all kinds of speculative (and oftentimes wrongheaded) head-games. "The ethical relationship," Levinas (1987, 56) writes, "is not grafted on to an antecedent relationship of cognition; it is a foundation and not a superstructure. ... It is then more *cognitive* than cognition itself, and all objectivity must participate in it."

This means that the order of precedence in moral decision-making can, and perhaps should be, reversed. Internal, substantive properties do not come first and then moral respect follows from this ontological fact. Instead, the morally significant properties—those ontological criteria that we assume ground moral respect—are what Slavoj Žižek (2008a, 209) terms "retroactively (presup)posited" as the result of and as justification for decisions made in the face of social interactions with others. In other words, we project the morally relevant properties onto or into those others who we have already decided to treat as being socially significant—those Others who are deemed to possess face, in Levinasian terminology. In social situations—in contending with the exteriority of others—we always and already decide between *who* counts as morally significant and *what* does not, and then retroactively justify these actions by "finding" the properties that we believe motivated this decision-making in the first place. Properties, therefore, are not the intrinsic *a priori* condition of possibility for moral standing. They are *a posteriori* products of extrinsic social interactions with and in the face of others.

Once again, this is not some heady theoretical formulation; it is practically the definition of machine intelligence. Although the phrase "artificial intelligence" is the product of an academic conference organized by John McCarthy at Dartmouth College in 1956, it is Alan Turing's 1950 paper and its "game of imitation," or what is now routinely called "the Turing Test," that defines and characterizes the field. Although Turing begins his essay by proposing to consider the question "Can machines think?" he immediately recognizes the difficulty with defining the subject "machine" and the property "think." For this reason, he proposes to pursue an alternative line of inquiry, one that can, as he describes it, be "expressed in relatively unambiguous words" (Turing 1999, 37). "The new form of the problem can be described in terms of a game which we call the 'imitation game.' It is played with three people, a man (A), a woman (B), and an interrogator (C) who may be of either sex. The interrogator stays in a room apart from the other two. The object of the game for the interrogator is to determine which of the other two is the man and which is the woman" (Turing 1999, 37). This determination is to be made on the basis of simple questions and answers. The interrogator (C) asks both the man (A) and the woman (B) various questions, and based on their responses (and only their responses) to these inquiries tries to discern whether the respondent is a man or a woman. "In order that tone of voice may not help the interrogator," Turing further stipulates, "the answers should be written, or better still, typewritten. The ideal arrangement is to have a teleprinter communicating between the two rooms" (Turing 1999, 37–38).

Turing then takes his thought experiment one step further. "We can now ask the question, 'What will happen when a machine takes the part of A in this game?' Will the interrogator decide wrongly as often when the game is played like this as he does when the game is played between a man and a woman? These questions replace our original, 'Can machines think?'" (Turing 1999, 38). In other words, if the man (A) in the game of imitation is replaced with a computer, would this device be able to respond to questions and simulate, to a reasonably high degree of accuracy, the activity of another person? If a computer is capable of successfully simulating a human being in communicative exchange to such an extent that the interrogator cannot tell whether she is interacting with a machine or another person, then that machine *should*, Turing concludes, be considered "intelligent." Or in Žižek's terms, if the machine effectively passes for another human person in communicative interactions, the property of intelligence would need to be "retroactively (presup)posited" for that entity, and this is done irrespective of the actual internal states or operations of the other, which are, according to the stipulations of Turing's test, unknown and hidden from view.

## 6.2  Applied (Levinasian) Philosophy

The advantage of applying Levinasian philosophy to the question of robot rights is that it provides an entirely different method for responding to the challenge not just of robots and other forms of autonomous technology but of the ways we have typically tried to address and decide things in the face of this very challenge. What is of fundamental importance is that this Levinasian alternative shifts the focus of the question and transforms the terms of the debate. Here it is no longer a matter of deciding, for example, whether robots can have rights or not, which is largely an ontological query concerned with the prior discovery of intrinsic and morally relevant properties. Instead, it involves making a decision concerning whether or not a robot—or any other entity for that matter—ought to have standing or moral/social status, which is an ethical question and one that is decided not on the basis of what things are but on how we relate and respond to them in actual social situations and circumstances. "The relational approach," as Coeckelbergh (2010a, 219) concludes, "suggests that we should not assume that there is a kind of moral backpack attached to the entity in question; instead, moral consideration is granted within a dynamic relation between humans and the entity under consideration." In this case, then, the actual practices of social beings in interaction with each other take precedence over the ontological properties of the individual entities or their different material implementations. This change in perspective provides for a number of important innovations that affect not just the ethical opportunities and challenges of robots but of moral philosophy itself.

### 6.2.1 The Face of the Robot

Looked at through the lens of Levinasian philosophy,[8] the question "Can or should robots have rights?" becomes something like "Can or should robots have face?" But this formulation is not quite accurate insofar as the verb "to have," as in "to have face" or "to have a face," has the tendency to turn "face" into a possession and a property that belongs to someone or something. The form of the question, therefore, risks redeploying the properties approach to deciding moral standing in the process of trying to articulate an alternative. A similar complication has been identified and addressed by Derrida (1978, 105) concerning the word "Other," which is "a substantive—and such it is classed by the dictionaries—but a substantive which is not, as usual, a species of noun: neither common noun, for it cannot take, as in the category of the other in general, the *heteron*, the definite article. Nor the plural," since implied in that kind of usage is *"property, rights: the property, the rights of Others"* (Derrida 1978, 105). Instead of asking "Can or Should robots *have* face?" or "Can or should robots be *the Other*?" we should rework the form of the question: "What does it take for a robot to supervene and be revealed as Other in the Levinasian sense?" This question, which recognizes that "alterity is a verb" no longer asks about "moral standing in a strict sense, since 'standing' suggests that there is an ontological platform onto which morality is mounted. It is therefore a more 'direct' ethical question: under what conditions can a robot—this particular robot that appears here before me—be included in the moral community?" (Coeckelbergh and Gunkel 2014, 723–24, slightly modified).

In order to respond to this other question (a question that is formulated otherwise and that is able to address others and other kinds of others), we need to consider not "robot" as a kind of generic ontological category but a specific instance of an encounter with a particular entity—one that challenges us to think otherwise. In July of 2014, for instance, the world got its first look at Jibo. Exactly who or what is Jibo? That is an interesting and important question. In a promotional video that was designed to raise capital investment through preorders, social robotics pioneer Cynthia Breazeal introduced Jibo with the following explanation: "This is your car. This is your house. This is your toothbrush. These are your things. But these [and the camera slowly zooms into a family photograph] are the *things* that matter. And somewhere in between is this guy. Introducing Jibo, the world's first family robot" (Jibo 2014). Whether explicitly recognized as such or not, this promotional video leverages Derrida's (2005, 80) distinction between *who* and *what*. On the side of *what* we have those things that are considered to be mere objects—our car, our house, and our toothbrush. According to the instrumental theory of technology, these things are mere instruments that do not have any independent moral status whatsoever (Lyotard 1984, 44). We might worry about the impact that the car's emissions have on the environment (or perhaps stated more precisely, on the health and well-being of the other human beings who share the planet with us), but the car itself is not a moral subject. On the other side there are, as the promotional video describes it "those things that matter." These things are not things, strictly speaking, but are the other persons who count as socially and morally significant Others. Unlike the car, the house, or the toothbrush, these Others have moral status and rights (i.e., privileges, claims, powers, and immunities). They can, in other words, be benefitted or harmed by our decisions and actions.

Jibo, we are told, occupies a place that is situated somewhere in between *what* are mere things and *who* really matters. Consequently Jibo is not just another instrument, like the automobile or toothbrush. But he/she/it (and again the choice of pronoun is not without consequence) is also not quite another member of the family pictured in the photograph. Jibo inhabits a place that is situated in between these two options. He/she/it occupies the in-between position Ihde (1990, 98) calls the "quasi-other" or what Prescott identifies as "liminal":

> Whilst most robots are currently little more than tools, we are entering an era where there *will* be new kinds of entities that combine some of the properties of machines and tools with psychological capacities that we had previously thought were reserved for complex biological organisms such as humans. Following Kang [2011, 17], the ontological status of robots might be best described as *liminal*—neither living quite in the same way as biological organisms, nor simply mechanical as with a traditional machine. The liminality of robots makes them both fascinating and inherently frightening, and a lightning rod for our wider fears about the dehumanising effects of technology. (Prescott 2017, 148)

This is, it should be noted, not unprecedented. We are already familiar with other entities who/that occupy a similar ambivalent social position, like the family dog. In fact animals, which since the time of Descartes's *bête-machine* have been the other of the machine (Gunkel 2012 and 2017b), provide a useful precedent for understanding the opportunities and challenges of sociable robots, like Jibo. Some animals, like the domestic pigs that are raised for food, occupy the position of *what*, being mere things that can be used and disposed of as we see fit. Other animals, like a pet dog, are closer (although not identical) to another person *who* counts as Other. They are named, occupy a place alongside us inside the house, and are considered by many to be "a member of the family" (Coeckelbergh and Gunkel 2014).

As we have seen, we typically theorize and justify the decision between the *what* and the *who* on the basis of intrinsic properties. This approach puts ontology before ethics, whereby what an entity is determines how it comes to be treated. But this method, for all its expediency, also has considerable difficulties (as previously described in chapter 3): (1) substantive problems with inconsistencies in the identification and selection of the qualifying property; (2) terminological troubles with the definition of the morally significant property; (3) epistemological complications with detecting and evaluating the presence of the property in another; and (4) moral concerns caused by the very effort to use this determination to justify extending moral standing to others. In fact, if we return to the example of animals, it seems difficult to justify differentiating between the pig, which is a thing we raise and slaughter for food and other raw materials, and the dog, who is considered to be (at least from the perspective of a contemporary European/North American cultural context, if not beyond) a member of the family, on the basis of ontological properties. In terms of all the usual criteria—consciousness, sentience, suffering, etc.—the pig and the dog seem to be (as far as we are capable of knowing and detecting) virtually indistinguishable. Our moral theory might dictate strict ontological criteria for inclusion and exclusion, but our everyday practices seem to operate otherwise, proving George Orwell's *Animal Farm* (1945, 118) correct: "All animals are equal, but some animals are more equal than others."

Alternative approaches to making these decisions, like that developed by Levinas, recognize that who is or can be Other is much more complicated. The dog, for instance, occupies the place of an Other (or at least Ihde's "quasi-other") who counts, while the pig is excluded as a mere thing, not because of differences in their intrinsic properties, but because of the way these entities have been situated in social relationships with us. One of these animals shares our home, is bestowed with a proper name, and is considered to have face, to use the Levinasian terminology. The other one does not. Social robots, like an animal, occupy an essentially undecidable or liminal position that is in between *who* and *what*. Whether Jibo or another kind of social robot is or is not an Other, therefore, is not something that will be decided in advance and on the basis of intrinsic properties; it will be negotiated and renegotiated again and again in the face of actual social circumstances and interactions. It will, in other words, be out of the actual social relationships we have with social robots that one will decide whether a particular robot counts or not. It is precisely for this reason that Prescott hesitates to dismiss or write-off the actual experiences of users. According to his analysis, what users of technology do in the face of the robot matters and needs to be taken seriously. "We should," Prescott (2017, 145) concludes, "take into account how people see robots, for instance, that they may feel themselves as having meaningful and valuable relationships with robots, or they may see robots as having important internal states, such as the capacity to suffer, despite them not having such capacities."

Jibo, and other social robots like this, are not science fiction. They are already or will soon be in our lives and in our homes.[9] And in the face of these socially situated and interactive entities, we are going to have to decide whether they are mere things like our car, our house, and our toothbrush; someone who matters like another member of the family; or something altogether different that is situated in between the one and the other as a kind of quasi-other. In whatever way this comes to be decided, however, these entities undoubtedly challenge our concept of ethics and the way we typically distinguish between *who* is to be considered Other and *what* is not. Although there are certainly good reasons to be concerned with how these technologies will be integrated into our lives and what the effect of that will be, this concern does not justify alarmist reactions and exclusions. We need, following the moral innovations of Levinas, to hold open the possibility that these devices might also implicate us in social relationships where they take on or are attributed face. At the very least, ethics obligates us—and it does so in advance of knowing anything at all about the inner workings and ontological status of these other kinds of entities—to hold open the possibility that they might become Other. Turkle (2011, 9), therefore, is right about one thing, we are and we should be "willing to seriously consider robots not only as pets but as potential friends, confidants, and even romantic partners." But this is not, as Turkle had initially suggested, a dangerous weakness or vulnerability to be avoided at all costs. It is ethics.

## 6.2.2 Ethics Beyond Rights

From one perspective, this Levinasian influenced, relational ethic might appear to be similar, at least functionally similar, to that developed in the second pair of modalities—those that distinguish between and do not derive *ought* from *is*. Kate Darling, in particular, advances a case for extending moral and legal considerations to social robots irrespective of what they are or can be. So what makes the Levinasian proposal different from the fourth modality (chapter 5)? The important difference is, in a word, *anthropomorphism*. For Darling, the reason we ought to treat robots well is because we perceive something of ourselves in them. Even if these traits and capabilities are not really there in the mechanism, we project them onto others—and other forms of otherness, like that represented by social robots—by way of anthropomorphism. For Darling, then, it is because the other looks and feels (to us) to be something like us that we are then obligated to extend to it some level of moral and legal consideration.

For Levinas this anthropomorphic operation is *the problem* insofar as it reduces the other to a modality of the same—turning what is other into an alter ego and mirrored projection of oneself. Levinas deliberately resists this gesture that already domesticates and even violates the alterity of others.[10] Ethics for Levinas is entirely otherwise: "The strangeness of the Other, his irreducibility to the I, to my thoughts and my possessions, is precisely accomplished as a calling into question of my spontaneity, as ethics" (Levinas 1969, 43). For Levinas, then, ethics is not based on "respect for others" but transpires in the face of the other and the interruption of *ipseity* that it produces. The principal moral gesture, therefore, is not the conferring or extending of rights to others as a kind of benevolent gesture or even an act of compassion but deciding how to respond to the Other who supervenes before me in such a way that always and already places me and my self-mastery in question. Levinasian philosophy, therefore, deliberately interrupts and resists the imposition of power that Birch (1995, 39) finds operative in all forms of rights discourse: "The nub of the problem with granting or extending rights to others ... is that it presupposes the existence and the maintenance of a position of power from which to do the granting." Whereas Darling (2012 and 2016a) is interested in "extending legal protection to social robots," Levinas provides a way to question the assumptions and consequences involved in this very gesture of power over others.

This alternative configuration, therefore, does not so much answer or respond to the question with which we began as it alters the terms of the inquiry itself. When one asks "Can or should robots have rights?" the form of the question already makes an assumption: namely, that rights are a kind of personal property or possession that an entity can have or should be bestowed with. Levinas, however, does not use the concepts or language of rights, and this is deliberate. "For rights," as Hillel Steiner (2006, 459) points out, "are essentially about who is owed what by whom." Considerations of rights, therefore, involve a subject (who) and an object (whom) (Sumner 2000, 289). But this gets ahead of things by making a prior decision about who or what can be the subject and object of rights. As Derrida (1978, 85) points out, Levinasian philosophy is directed against and works to articulate what is anterior to the "primacy of the subject-object correlation." For this reason, the Other is not, strictly speaking, either the subject of rights (i.e., one or more of the Hohfeldian incidents: privilege, claim, power, and/or immunity) or subjected to rights (as typically formulated in terms of either the Will or Interest Theory). As Coeckelbergh (2016c, 190) succinctly describes it, otherness is not a noun: "alterity is a verb." The Other (despite this grammatical construction) is not a substantive subject position that can be declared in advance of interactions with others. It is an action or a happening that needs to be responded to in the act of specific social challenges and interactions. "It is," as Coeckelbergh (2016c, 190) explains, mobilizing Heideggerian terminology, "about experiencing (*Erfahren*) and something happening to you, which can assume the dimension of a meeting or en-counter (*Wiederfahren*)."

Levinasian ethics, then, is not reducible to the Hohfeldian incidents of privileges, claims, powers, and/or immunities; nor is it limited to and determined by the Will Theory or Interest Theory of rights. It articulates a mode of thinking that is prior and anterior to these standard determinations. "For the condition for," Levinas (1987, 123) explains, "or the unconditionality of, the self does not begin in the autoaffection of a sovereign ego that would be, after the event, 'compassionate' for another. Quite the contrary: the uniqueness of the responsible ego is possible only in being obsessed by another, in the trauma suffered prior to any auto-identification, in an unrepresentable before." The self or the ego, as Levinas describes it, does not constitute some preexisting self-assured condition that is situated before and is the cause of the subsequent relationship with another. It does not (yet) take the form of empathy or compassion or even respect for the rights (i.e., a privilege, a claim, a power, and/or an immunity) of the other. Rather it becomes what it is as a byproduct of an uncontrolled and incomprehensible exposure to the face of the Other that takes place prior to any formulation of the self in terms of agency.

Likewise, the Other is not comprehended as a "moral patient" who is constituted as the recipient of the agent's actions and whose interests and rights would need to be identified, taken into account, and duly respected. Instead, the absolute and irreducible exposure to the Other is something that is anterior and exterior to these determinations, not only remaining beyond the range of their conceptual grasp and regulation but also making possible and ordering the antagonistic structure that subsequently comes to characterize the difference that distinguishes the self from its others, the agent from the patient, and the subject from the object of rights. In other words, for Levinas at least, prior decisions about subject and object or moral agent and moral patient do not first establish the terms and conditions of any and all possible encounters that the self might have with others and with other forms of otherness. It is the other way around. The Other first confronts, calls upon, and interrupts self-involvement and in the process determines the terms and conditions by which the standard roles of the subject and object of rights come to be articulated and assigned. An ethics without (or anterior to) rights is an ethics that is open to the moral situation of others.

## 6.3 Complications, Difficulties, and Potential Problems

This is not to say that Levinas's innovations do not have their own challenges and problems. And criticisms of Levinasian philosophy are not in short supply. For our purposes, there are at least two important difficulties that complicate the application of Levinas's work to robots.

### 6.3.1 Anthropocentrism
Although Levinas effectively dodges and disarms the complications of anthropomorphism (as mobilized in the work of Kate Darling and others), his philosophy, for all its promise to organize a mode of thinking that is organized and oriented otherwise, has been and remains inescapably anthropocentric. For this reason, applying his work to robots, let alone

nonhuman animals or the environment, can only occur by way of a deliberate violation of the proper limits that he himself had established for his own brand of philosophical thinking. Whatever the import of his unique contribution, "Other" in Levinas is and remains unapologetically human. Although he is not the first to identify it, Jeffrey Nealon (1998, 71) provides what is perhaps one of the most succinct descriptions of the problem: "In thematizing response solely in terms of the human face and voice, it would seem that Levinas leaves untouched the oldest and perhaps most sinister unexamined privilege of the same: *anthropos* [ἄνθρωπος] and only *anthropos*, has *logos* [λόγος]; and as such, *anthropos* responds not to the barbarous or the inanimate, but only to those who qualify for the privilege of 'humanity,' only those deemed to possess a face, only to those recognized to be living in the *logos*." For Levinas, as for many of those who follow in the wake of his influence, Other has been exclusively operationalized as another human subject. If, as Levinas argues, ethics precedes ontology, then in Levinas's own work anthropology and a certain brand of humanism still precede ethics.

There is considerable debate in the subsequent literature concerning this anthropocentrism and its significance. Derrida, for his part, finds it a disturbing residue of humanist thinking and therefore reason for considerable concern: "In looking at the gaze of the other, Levinas says, one must forget the color of his eyes, in other words see the gaze, the face that gazes before seeing the visible eyes of the other. But when he reminds us that the 'best way of meeting the Other is not even to notice the color of his eyes,' he is speaking of man, of one's fellow as man, kindred, brother; he thinks of the other man and this, for us, will later be revealed as a matter for serious concern" (Derrida 2008, 12). What truly "concerns" Derrida is not just the way this anthropocentrism (which in this particular context also exhibits an exclusive gendered configuration that is part and parcel of the humanist tradition) restricts Levinas's philosophical innovations but the manner by which it already makes exclusive decisions about the (im)possibility of an animal other. Richard Cohen, by contrast, endeavors to give this "problem" a positive interpretation in his "Introduction" to one of Levinas's books: "The three chapters of *Humanism of the Other* each defend humanism—the world view founded on the belief in the irreducible dignity of humans, a belief in the efficacy and worth of human freedom and hence also of human responsibility" (Cohen 2003, ix). For Cohen, however, this version of humanism is significantly different; it consists in a radical thinking of the "humanity of the human" as the unique site of ethics: "From beginning to end, Levinas's thought is a humanism of the other. The distinctive moment of Levinas's philosophy transcends its articulation but is nevertheless not difficult to discern: the superlative moral priority of the other person. It proposes a conception of the 'humanity of the human,' the 'subjectivity of the subject,' according to which being 'for-the-other' takes precedence over, is better than being for-itself. Ethics conceived as a metaphysical anthropology is therefore nothing less than 'first philosophy'" (Cohen 2003, xxvi).

Consequently, utilizing Levasian thought for the purposes of addressing the moral/legal status of robots requires fighting against and struggling to break free from the gravitational pull of Levinas's own anthropocentrism. Fortunately this application—this effort to read Levinas in excess of Levinas—is not unprecedented. There has, in fact, been some impressive efforts at "radicalizing Levinas" (Atterton and Calarco 2010), "broadening Levinas's sense of the Other" (Davy 2007, 41), and elaborating Levinasian philosophy in "the face of things" (Benso 2001). "Although Levinas himself," as Calarco (2008, 55) points out: "is for the most part unabashedly and dogmatically anthropocentric, the underlying logic of his thought permits no such anthropocentrism. When read rigorously, the logic of Levinas's account of ethics does not allow for either of these two claims. In fact ... Levinas's ethical philosophy is, or at least should be, committed to a notion of universal ethical consideration, that is, an agnostic form of ethical consideration that has no a priori constraints or boundaries." In proposing this alternative reading of the Levinasian tradition, Calarco interprets Levinas against himself, arguing that the logic of Levinas's account is in fact richer and more radical than the limited interpretation the philosopher had initially provided for it. This means, of course, that we would be obligated to consider all kinds of others as Other, including other human persons, animals, the natural environment, artifacts, technologies, and robots. An "altruism" that tries to limit in advance who can or should be Other would not be, strictly speaking, altruistic.

Unfortunately even Calarco's "radicalization of Levinas" does not go far enough. His representative list of previously excluded others, which includes "'lower' animals, insects, dirt, hair, fingernails, and ecosystems" (Calarco 2008, 71), may be more responsive to other forms of otherness, but it also makes exclusive decisions. And what is noticeably absent from his list are artifacts, technologies, and robots. The same may be said for Barbara Jane Davy, who also seeks to generalize Levinas's philosophy beyond its limited anthropocentric restrictions. Like Calarco, Davy's list of other kinds of others is broader, but it is still lacking anything artificial or technological:

> In broadening Levinas's sense of the Other I aim to include not just humans and other animals, but any Other. While for Levinas the Other is always assumed to be a human being, I take his phenomenological understanding of the Other beyond categories such as human, animal, plant, rock, wind, or body of water. In Levinasian ethics, the Other is met as a person rather than thematized or interpreted through categories. The Other interrupts one's thematization of everything into one's own view of the world. I contend that other sorts of persons can also interrupt one's thematization of the world in this manner. (Davy 2007, 41)

Despite efforts to develop Levinas's philosophy beyond its own inherent limitations to an anthropocentric frame and point of reference, these subsequent efforts continue to institute marginalizations of any kind of technological other. In the face of all kinds of other things that can in principle take on face—animals, plants, rocks, waters, and even fingernails—robots remain faceless and otherwise than Other, or beyond ethics.

### 6.3.2 Relativism and Other Difficulties

For all its opportunities, the extension of Levinasian philosophy to other forms of otherness risks exposure to the charge of moral relativism—"the claim that no universally valid beliefs or values exist" (Ess 1996, 204) or "that beliefs, norms, practices, frameworks, etc., are legitimate in relation to a specific culture" (Ess 2009, 21). To put it rather bluntly, if moral status is "relational" (Coeckelbergh's term) and open to different decisions concerning others made at different times for different reasons, are we not at risk of affirming an extreme form of moral relativism? The answer to this question depends on how one defines the term "relativism." As I have argued elsewhere (Gunkel 2010 and 2012) "relative," which has an entirely different pedigree in a discipline like physics, need not be construed negatively and decried, as Žižek (2000, 79 and 2006, 281) has often done, as the epitome of postmodern multiculturalism run amok, or, as Robert Scott (1976, 264) characterizes it, as "a standardless society, or at least a maze of differing standards, and thus a cacophony of disparate, and likely selfish, interests." Instead, it may be the case that, following Scott, "relativism can indicate circumstances in which standards have to be established cooperatively and renewed repeatedly" (Scott 1976, 264). This means that one can remain critical of "moral relativism," in the usual dogmatic sense of the phrase, while being open and receptive to the fact that moral standards—like many social conventions and legal statues—are socially constructed formations that are subject to and the subject of difference.

Charles Ess (2009, 21) calls this alternative "ethical pluralism." "Pluralism stands as a third possibility—one that is something of a middle ground between absolutism and relativism. ... Ethical pluralism requires us to think in a 'both/and' sort of way, as it conjoins both shared norms and their diverse interpretations and applications in different cultures, times, and places" (Ess 2009, 21–22). Likewise Floridi (2013, 32) advocates a *"pluralism without endorsing relativism,"* calling this "middle ground" *relationalism*: "When I criticize a position as *relativistic*, or when I object to *relativism*, I do not mean to equate such positions to non-absolutist, as if there were only two alternatives, e.g. as if either moral values were absolute or relative, or truths were either absolute or relative. The method of abstraction enables one to avoid exactly such a false dichotomy, by showing that subjectivist positions, for example, need not be relativistic, but only relational" (Floridi 2013, 32). And Žižek, for his part, advances a similar position, which he formulates not in terms of ethics but epistemology: "At the level of positive knowledge it is, of course, never possible to (be sure that we have) attain(ed) the truth—one can only endlessly approach it, because language is ultimately self-referential, there is no way to draw a definitive line of separation between sophism, sophistic exercises, and Truth itself (this is Plato's problem). Lacan's wager is here the Pascalean one: the wager of Truth. But how? Not by running after 'objective' truth, but by holding onto the truth about the position from which one speaks" (Žižek 2008b, 3). Like Floridi, Žižek recognizes that truth is neither absolute (always and everywhere one and the same) nor completely relative (such that "anything goes"); it is always formulated and operationalized from a particular position of enunciation, or what Floridi calls "level of abstraction," which is dynamic and alterable.

In endorsing this kind of "ethical pluralism" or "theory of moral relativity" and following through on it to the end, what one gets is not a situation where anything goes and "everything is permitted" (Camus 1983, 67). Instead, what is obtained is a kind of ethical thinking that is more dynamic and responsive to the changing social situations and circumstances of the twenty-first century. Although such a formulation sounds workable in theory, it does run into considerable problems when deployed and put into practice. In particular, if any and all things become capable of taking on face—animals, plants, rocks, and even robots—does this also indicate that something might not, for one reason or another, manifest face and therefore come to be treated as a mere tool or instrument for my use and enjoyment? Is it possible that what is to be gained for others, including the robot—namely, that they might supervene before us and possess face—could in fact work in the opposite direction and *deface* another (another human individual, for instance), who for one reason or another would not achieve the same kind of moral status? If face is not a kind of substantive property of Others, but a kind of event or occurrence of alterity, does the "relational turn" run the risk, as noted by Gerdes (2015, 274), that we might lose something valuable, that "our human-human relations may be obscured by human-robot relations?"

This is a much more practically oriented question and one that cannot be answered by simply deferring to the concept of ethical pluralism or a theory of moral relativity. One possible mode of response—what we might call the conservative version—reaffirms human exceptionalism in the face of all others. This is the position occupied and advocated by Kathleen Richardson:

> But if the machine can become another, what does it say for how robotic and AI scientists conceptualise 'relationship'? Is relationship instrumental? Is relationship mutual and reciprocal? There are many different kinds of relationships that people have. Our market economies structure work so that encounters between people in the work sphere take on a character of formal interactions. ... These formal interactions characterize a huge proportion of our lived experience. Some philosophers have characterised these types of relations between persons as "instrumental." If these kinds of relations are "instrumental" does that make people in the situations "instruments" or "tools?" Humans are never tools or instruments, even if relations between people take on a formal character. When we meet a cashier at a checkout or restaurant service staff, they have not stopped being human just because they are only expressing themselves formally in the given situation. People do not stop being human and turn into instruments when they enter the working environment, and then switch back to being human in the private sphere. In every encounter we meet each other as persons, members of a common humanity. (Richardson 2017, 1)

Richardson's assertion that "humans are never tools or instruments" not only ignores the fact that for a good part of human history (and even now) humans beings have treated other human individuals and communities of individuals as tools or instruments (i.e., slaves) but also takes the rhetorical form of a dogmatic and absolute proclamation, deploying totalizing words like "never" and "in every encounter." In other words, Richardson's response to the potential hazards of relationalism or relativism is to retreat to an absolutist position through a kind of totalitarian assertion concerning the truth of things. But it is precisely this gesture and its legacy—this assertion of an unquestioned human exceptionalism and the reduction of difference to the same by way of a common underlying "humanity" that is always defined and defended from some particular position of power—that is in question.

A different way to respond to this question is to recognize, as Anne Foerst suggests (mobilizing the terminology of personhood that had already appeared in Richardson's assertion), that otherness is not (not in actual practices, at least) unlimited or absolute: "Each of us only assigns personhood to a very few people. The ethical stance is always that we have to assign personhood to everyone, but in reality we don't. We don't care about a million people dying in China of an earthquake, ultimately, in an emotional way. We try to, but we can't really, because we don't share the same physical space. It might be much more important for us if our dog is sick" (Benford and Malartre 2007, 163). This statement is perhaps more honest about the way that moral decision-making and the occurrence of otherness actually transpires. Instead of declaring an absolutist claim to a kind of totality (Levinas's word), it remains open to particular configurations of otherness that is more mobile, flexible, and context-dependent. But there remains, as Gerdes (2015) recognizes, something in this formulation that is abrasive to our moral intuitions. This may be due to the fact that this Levinasian inspired ethic does not make a singular and absolute decision about otherness that stands once and for all, such that there is one determination concerning others that decides everything for all time. The encounter with Others—the occurrence of face in the face of the Other—is something that happens in time and needs to be negotiated and renegotiated again and again. The work of ethics is, therefore, inexhaustible. It is an ongoing and interminable responsibility requiring that one respond and take responsibility for how one responds. Is this way of thinking and doing ethics without risk? Not at all. But the risk is itself the site of ethics and the challenge that one must face in all interactions with others, whether human, animal, or otherwise.

## 6.4 Summary

"Every philosophy," Benso (2000, 136) writes in a comprehensive gesture that performs precisely what it seeks to address, "is a quest for wholeness." This objective, she argues, has been typically targeted in one of two ways: "Traditional Western thought has pursued wholeness by means of reduction, integration, systematization of all its parts. Totality has replaced wholeness, and the result is totalitarianism from which what is truly other escapes, revealing the deficiencies and fallacies of the attempted system" (Benso 2000, 136). This is precisely the kind of violent philosophizing that Levinas (1969) identifies under the term "totality," and it includes the efforts of both standard agent-oriented approaches (i.e., consequentialism, deontologism, virtue ethics, etc.) and non-standard patient-oriented approaches (i.e., animal rights, bioethics, and information ethics). The alternative to these totalizing transactions is a philosophy that is oriented otherwise, like that proposed by Levinas. This other approach, however, "must do so by moving not from the same, but from the other, and not only the Other, but also the other of the Other, and, if that is the case, the other of the other of the Other. In this *must*, it must also be aware of the inescapable injustice embedded in any formulation of the other" (Benso 2000, 136). And this "injustice" that Benso identifies and calls-out is evident not only in Levinas's exclusive anthropocentrism but also in the way that those who seek to redress this "humanism of the other" continue to exclude or marginalize technology in general and robots in particular.

For these reasons, the question concerning robot rights does not end with a single definitive answer—a simple and direct "yes" or "no." But this outcome is not, we can say following the argumentative strategy of Daniel Dennett's "Why you Cannot Make a Computer that Feels Pain" (1978 and 1998), necessarily a product of some inherent or essential deficiency with the technology. Instead it is a result of the fact that the discourse of rights as well as the methods by which moral standing has been typically questioned and resolved already deploy and rely on questionable constructions and logics. "Whether or not it is acceptable to grant rights to some robots," as Coeckelbergh (2010a, 219) concludes, "reflection on the development of artificially intelligent robots reveals significant problems with our existing justifications of moral consideration." The question concerning the rights of robots, therefore, is not simply a matter of asking about the extension of moral consideration to one more historically excluded other, which would, in effect, leave the existing mechanisms of moral philosophy in place, fully operational, and unchallenged. Instead, the question of robot rights (assuming that it is desirable to retain this particular vocabulary) makes a fundamental claim on ethics, requiring us to rethink the systems of moral considerability all the way down. This is, as Levinas (1981, 20) explains, the necessarily "interminable methodological movement of philosophy" that continually struggles against accepted norms and practices in an effort to think otherwise—not just differently, but in ways that are responsive to and responsible for others.

Consequently, this book ends not as one might have initially expected—that is, by accumulating evidence or arguments in favor of permitting robots, a particular kind of robot, or even just one representative robot, entry into the community of moral and/or legal subjects. Instead, it concludes with a fundamental challenge to ethics and the way moral philosophy has typically defined, decided, and defended the question of the standing and status of others. Although ending in this way—in effect, responding to a question with yet another question—is commonly considered bad form, this is not necessarily the case. This is because questioning is definitive of the philosophical enterprise. "I am," Dennett (1996, vii) writes, "a philosopher and not a scientist, and we philosophers are better at questions than answers. I haven't begun by insulting myself and my discipline, in spite of first appearances. Finding better questions to ask, and breaking old habits and traditions of asking, is a very difficult part of the grand human project of understanding ourselves and our world." The objective of *Robot Rights*, therefore, has not been to advocate on behalf of a cause, to advance some definitive declaration, or to offer up a preponderance of evidence to prove one or the other side in the existing debates. The goal has been to ask about the very means by which we have gone about trying to articulate and formulate this problem and its investigation. The issue, then, is not only investigating whether or not robots can and should have rights, but also, and perhaps more importantly, exposing how these questions concerning rights have been configured and how these configurations already accommodate and/or marginalize others. Or to put it in terms devised by Žižek (2006b, 137), the question "Can and should robots have rights?" might already be questionable insofar as the very form of the inquiry—the very way we have perceived and defined the problem—"is already an obstacle to its solution."

For this reason, the examination of robot rights is not just one more version or iteration of an applied moral philosophy; its investigation opens up and releases a thorough and profound challenge to what is called "ethics." And answering this call will be the task of thinking of and for the future. In other words, if the problem from the beginning has been the fact that the question of robot rights is and remains "unthinkable" (Levy 2005, 393)—or if not entirely unthinkable then at least so difficult to think[11] that it occupies a marginal position in the efforts of roboethics, robot ethics, machine ethics, robophilosophy, etc.—then it will have been sufficient for us that the book has made it possible to think the unthought (to use that distinctly Heideggerian formulation) or (if one would prefer something less continental and more analytic) to render tractable the question "Can and should robot have rights?" Where we go from here, or what can and should be thought or done from this point forward, remains a question that will need to be answered in the face of the robot.

# Notes

## Introduction

1. The works that are cited in this paragraph are only representative and their enumeration here makes no pretense to be exhaustive. A decade ago this would not have been the case. When I first began publishing in the field of artificial intelligence, robotics, and ethics (Gunkel 2007), it was entirely feasible to attempt a comprehensive review of the existing literature. Even by the time of *The Machine Question* (Gunkel 2012), such an effort would not have been difficult. But, shortly thereafter, the number of works available and in circulation exploded, making such efforts virtually impractical for standalone human processing.

2. For a detailed consideration of the concepts of agency and patiency in moral philosophy, see Hajdin (1994), Floridi and Sanders (2004), and Gunkel (2012).

3. Targeting and interrogating texts, and not individuals, has an important added benefit. In conducting his inquiries in the way he does, Socrates angered his fellow citizens by demonstrating that they did not know what they had professed to know. In response to this, the citizens of Athens charged Socrates with corruption, put him on trial, and eventually handed down a death sentence. Circumventing this kind of outcome is something that is entirely welcomed.

4. I have now used this particular formulation in a number of different places, so much so that it could be considered a kind of trademark for a research program. The phrase, as explained elsewhere (Gunkel 2007), indicates: 1) a way of thinking that is able to address and respond to others (and other forms of otherness); and 2) other ways of thinking about others and otherness that are significantly different from the usual way of proceeding.

5. For a detailed and thorough characterization of the deconstruction, see: the Appendix, "Deconstruction for Dummies," to *Hacking Cyberspace* (Gunkel 2001, 201–5); the first chapter, "Critique of Digital Reason," in *Thinking Otherwise: Philosophy, Communication, Technology* (Gunkel 2007, 11–43); and the introductions to *The Machine Question: Critical Perspectives on AI, Robots and Ethics* (Gunkel 2012) and

*Gaming the System: Deconstructing Video Games, Game Studies and Virtual Worlds* (Gunkel 2018).

6. Here is how Searle (1980, 417) initially defined these two varieties of AI: "I find it useful to distinguish what I will call 'strong' AI from 'weak' or 'cautious' AI (Artificial Intelligence). According to weak AI, the principal value of the computer in the study of the mind is that it gives us a very powerful tool. For example, it enables us to formulate and test hypotheses in a more rigorous and precise fashion. But according to strong AI, the computer is not merely a tool in the study of the mind; rather, the appropriately programmed computer really is a mind, in the sense that computers given the right programs can be literally said to understand and have other cognitive states." Subsequent usage of the term "weak AI" has (for better or worse) deviated from Searle's rather precise philosophical characterization. In much of the current literature on the subject, "weak AI" is now (mis)used as another name for "narrow AI," which identifies smart applications that do not reproduce human-grade sentience or consciousness (the "strong" variety of AI) but are confined to single and narrowly defined tasks (i.e., expert systems, recommendation algorithms, facial recognition applications, speech dialogue systems, machine translation, etc.). Whereas "strong AI" is still arguably situated in the realm of science fiction, "weak or 'narrow' AI," as Ryan Calo (2011, 1) explains, "is a present-day reality." In the context of this book, I use the terms "weak AI" and "strong AI" in the sense that was initially introduced and developed by Searle.

## Chapter 1

1. A similar point was made during an interview published on the Science Museum website:

**Interviewer:** Our view of robots is shaped by books and film—do you think this is helpful or misleading?
**Alan Winfield:** Good question! I think robots in science fiction are both helpful and misleading. Helpful because many roboticists, myself included, were inspired by science fiction, and also because SF provides us with some great examples of "thought experiments" for what future robots might perhaps be like—think of the movie AI, or Data from Star Trek. (Of course there are some terrible examples as well!) But robots in science fiction are misleading too. They have created an expectation of what robots are (or should be) like that means that many people are disappointed by real-world robots. This is a great shame because real-world robots are—in many ways—much more exciting than the fantasy robots in the movies. And the misleading impression of robots from SF makes being a roboticist harder, because we sometimes have to explain why robotics has "failed"—which of course it hasn't! (Winfield 2011b)

2. This extension is not without considerable debate and discursive effort in the subsequent literature. As L. W. Sumner (1987, 20) has pointed out: "Hohfeld was concerned solely with legal conceptions or, as he also called them, jural relations, and thus he was concerned solely with legal rights. By contrast, at this stage of our inquiry we are seeking an account of the internal structure not just of legal rights but of all forms of conventional rights. In order to construct this account we will

have to expand Hohfeld's narrow focus on rights within a municipal legal system to include rights with any conventional rule system, legal or non-legal."

3. Minsky's essay is itself a strange item. It is a kind of science fiction written in order to question whether human beings deserve rights: "Two interstellar aliens have come to assess the life-forms of Earth. The human life-forms will be entitled to rights—if the aliens can conclude that they think. Most such decisions are easy to make—but this case is unusual" (Minsky 1993).

4. The reasons behind these reactions are examined in my previous book, *The Machine Question* (Gunkel 2012), which identifies and documents the ontological, metaphysical, epistemological, and moral mechanisms of this exclusivity.

5. The response from the general public was no better as Seo-Young Chu (2010, 215) documents by way of user comments:

Most humans would rather dismiss such claims as frivolous and insubstantial than acknowledge them as part of reality. "This is a silly concept," one commentator remarked in response to the 2006 study released by the British government: "Granting a robot rights would be akin to granting the right front tire on my car rights" (Bryansix). Another commentator expresses his resistance in harsher language: "Robot rights are a joke and so are the morons that even consider them" (GoldFish). A third commentator voices his objection to robot rights in the most violent way possible: "If I buy the robot, I should be able to do whatever the hell I want to it. Including beatings, burnings, disabling, beheading" (Keep Robots Slaves). From amused skepticism to disgust and vehement hostility, such attitudes of mental rejection are symptomatic of the cognitively estranging nature of robot rights.

6. It may, of course, be objected that this interpretation simply makes too much out of what is nothing more than editorial exigency. When I asked Patrick Lin about it (via a personal email from August 2017), he offered two explanations for this perceived difference in the two versions of *Robot Ethics*: (1) "I wouldn't read too much into what we call our sections—that doesn't really track any perceived trend or delimit our areas of interest." And (2) the "book *does* engage the question of rights in several chapters ... it just made more sense to organize them topically in a different way." If you track the term "rights" as it occurs in the second version, however, there is just one essay that gives passing consideration to the rights of robots by way of its introductory material (Darling 2017). The remaining contributions, when they do take up and engage the question of rights, focus attention on the rights of human individuals and communities. Even the consideration of robot personhood, which Sparrow (2012) had developed in the first version, is now limited to matters of robot liability (White and Baum 2017, 70).

7. The logic of this operation was already formulated in Plato's *Timaeus*. The dialogue begins with a kind of roll call where Socrates accounts for who is present (or not) by way of counting: "One, two, three,—but where, my dear Timaeus, is the fourth of our guests of yesterday, the hosts of today?" (Plato 1981, 17a). In enumerating the individuals present before him, Socrates notices a difference between the number of people that had been present during the previous day's discussion (pre-

sumably the *Republic*, insofar as the *Timeaus* is situated as its sequel), and the number of participants who are currently present at the beginning of the current discussion. In the process of comparing what had been yesterday to what is today, Socrates identifies an absence. There is someone missing, someone who was supposed to be there but now is not. This absence, or lack of presence, supervenes and takes place insofar as someone is not there and in their place. Consequently the "missing fourth" of the *Timaeus* is available and only comes to presence by way of a taking account of this difference. The absence of the one who should have been present only comes to presence by way of this trace of their withdrawal.

8. Freitas's text appears under three different monikers. The original title of the "End Note" that was published in *Student Lawyer* was "Can the Wheels of Justice Turn for our Friends in the Mechanical Kingdom? Don't Laugh." The title that is listed in the journal's table of contents, however, is different: "When R2-D2 Needs a Lawyer: Remember, You Read it Here First." Furthermore a subsequent self-archived version of the document bears a third designation: "The Legal Rights of Robots." Citations of the work typically utilize and make reference to this third title.

9. This attention to "abuse" may be a product of the venue in which Whitby's essay initially appears, a special issue of *Computers and Interaction* addressing "the abuse and misuse of social agents." Although the term "abuse" might give some readers pause, insofar as the word might be perceived as being too anthropomorphic and not applicable to technological objects, it is explained and justified in the editors' introduction.

> We recognize that the title of this special issue, Abuse and Misuse of Social Agents, may annoy or disappoint some readers because of its metaphorical implications. Clearly, the word abuse, when defined as the intent to cause suffering, cannot be applied to inanimate objects, as they are incapable of recognizing, experiencing, or understanding suffering. ... Although we agree with the above statement, we do believe that the metaphor of social agency by itself tends to promote technology to a level that cannot escape the possibility of abuse. If it makes sense to say users are being *polite* in their responses to a conversational agent when obeying the social rules of politeness, then why would it not make equal sense to say a user is being abusive in his or her responses when using language that, if applied to people, would be considered abusive? (Brahnam and De Angeli 2008, 288)

10. Although Bensoussan and Bensoussan do not analyze this right into its Hohfeldian incidents, it is most likely the case that it would, like the right to property that was considered above, be composed of a compound of first- and second-order incidents.

11. The issue of suffering has the effect, Derrida (2008, 27) argues, of "changing the very form of the question regarding the animal": "Thus the question will not be to know whether animals are of the type *zoon logon echon* [ζωον λόγον ἔχον], whether they can speak or reason thanks to that *capacity* or that *attribute* of the *logos* [λόγος], the can-have of the *logos*, the aptitude for the logos (and logocentrism is first of all a thesis regarding the animal, the animal deprived of the logos, deprived of the *can-*

*have-the-logos*: this is the thesis, position, or presumption maintained from Aristotle to Heidegger, from Descartes to Kant, Levinas, and Lacan). The first and decisive question would be rather to know whether animals can suffer" (Derrida 2008, 27). For more on this shift in the question concerning the moral standing of others, see the final chapter and Gunkel (2012).

12. The use of words and terminology connected to vision is neither accidental nor merely metaphorical. The English word "theory," as we are often reminded, is derived from the ancient Greek θεωρία [theōría], which denotes the act of contemplation or a way of looking at something. A theory, therefore, like the frame of a camera, always enables something to be seen by including it within the field of vision, but it also and necessarily excludes other things outside the edge of its frame.

## Chapter 2

1. A recent addition (2013) to the FAQ published on the *Chicago Manual of Style* website addresses the decision in the context of a grammatical question concerning pronouns:

Q. When referring to a zombie, should I use the relative pronoun *who* (which would refer to a person) or *that* (since, technically, the zombie is no longer living)? Essentially, does a zombie cease to become a "person" in the grammatical sense?
A. Let's assume this is a serious question, in which case you, as the writer, get to decide just how much humanity (if any) and grammatical sense you wish to invest in said zombie. That will guide your choice of *who* or *that*.

2. Asking about things appears to be a rather pointless, if not remedial, exercise. We all know what things are; we deal with them every day. But, as Martin Heidegger (1962) pointed out, it is this immediacy and proximity that is precisely the problem. It is our closeness to things that make us miss what comprises their particular ontological character as things. Marshall McLuhan and Quentin Fiore (2001, 175) cleverly express it this way: "one thing about which fish know exactly nothing is water." Like fish that cannot perceive the water in which they live and operate, we are, Heidegger suggests, often unable to see the things that are closest to us and comprise the very milieu of our everyday existence. In response to this, Heidegger commits considerable effort to investigating what things are and why things seem to be more difficult (and interesting) than they initially appear to be. In fact, "the question of things," is one of the principal concerns and an organizing principle of Heidegger's ontological project (Benso 2000, 59), and this interest in things begins right at the beginning of his 1927 magnum opus, *Being and Time*: "The Greeks had an appropriate term for 'Things': πράγματα [pragmata]—that is to say, that which one has to do with in one's concernful dealings (πραξις). But ontologically, the specific 'pragmatic' character of the πράγματα is just what the Greeks left in obscurity; they thought of these 'proximally' as 'mere Things'. We shall call those entities which we encounter in concern 'equipment' [*Zeug*]" (Heidegger 1962, 96–97). According to Heidegger's analysis, things are not, at least not initially, experienced as objects out there in the

world. They are always pragmatically situated and characterized in terms of our involvements and interactions with the world in which we live. For this reason, things are first and foremost revealed as "equipment," which are useful for our endeavors and objectives. "The ontological status or the kind of being that belongs to such equipment," Heidegger (1962, 98) explains, "is primarily exhibited as 'ready-to-hand' or *Zuhandenheit*, meaning that some-thing becomes what it is or acquires its properly 'thingly character' when we use it for some particular purpose." According to Heidegger, then, the fundamental ontological status, or mode of being, that belongs to things is primarily exhibited as "ready-to-hand," meaning that some-thing becomes what it is or acquires its properly "thingly character" in coming to be put to use for some particular purpose. A hammer, one of Heidegger's principal examples, is for building a house to shelter us from the elements; a pen is for writing a book like this one; a shoe is designed to support the activity of walking. Everything is what it is in having a "for which" or a destination to which it is always and already referred. Everything therefore is primarily revealed as being a tool or an instrument that is useful for our purposes, needs, and projects. For a more detailed examination of Heidegger's philosophical consideration of things, see Benso (2000) and Gunkel and Taylor (2014).

3. The privileged example in these arguments is always a toaster. Take, for example, Wesley Smith's (2015) reply to the mere idea of robot rights: "My response to this is simple. Machines have no dignity and no rights, which properly belong exclusively to the human realm. Moreover, AI contraptions would only mimic sentience. As inanimate objects, AI contrivances could no more be 'harmed' (as distinguished from damaged) than a toaster." Toasters also appear as the principal example in the work of Kate Darling (2016, 216) and the Kurzgesagt–In a Nutshell (2017) video "Do Robots Deserve Rights? What if Machines Become Conscious?"—which begins with the following thought experiment:

Imagine a future where your toaster anticipates what kind of toast you want. During the day, it scans the Internet for new and exciting types of toast. Maybe it asks you about your day and wants to chat about new achievements in toast technology. At what level would it become a person? At which point will you ask yourself if your toaster has feelings? If it did, would unplugging it be murder? And would you still own it? Will we one day be forced to give our machines rights?

For this reason, Ron Moore's reboot of *Battlestar Galactica* (2004–2009) was remarkably insightful on this account. In *BSG* there are two types of robots, or Cylons: the utilitarian centurions, which are chrome-plated machine-looking mechanisms, and the "skin-jobs," which are artifacts that are ostensibly indistinguishable from real human beings. The former are called "toasters" by the human colonists—a term the skin-job Cylons find offensive and racist—and can be shot and killed with little hesitation insofar as they are just instruments or machines. Consequently, using the example of the toaster—instrumentalizing this particular instrument to explain the ins and outs of the instrumental theory—is not an accident.

4. In the revision (Boden et al. 2017), the phrase "experts" is replaced by "a group," such that the opening line of "The Principles" reads: "In September 2010, a group drawn from the worlds of technology, industry, the arts, law and social sciences met at the joint EPSRC and AHRC Robotics Retreat to discuss robotics, its applications in the real world and the huge promise robotics offers to society" (Boden et al. 2017, 124).

5. Andrea Bertolini (2013, 224) advances a similar argument in a consideration of what he calls "Strong Autonomy":

> As is well known, the Kantian categorical imperative forbids that a counterpart be reduced to a pure means to the actor's own end; the relationship between *alter* and *ego* needs to be construed so as to place both subjects on the same level, and thus the second ought to be able to pursue its own desires. The artificial agent instead would still be created for a given purpose and this "is a status that cannot be undone by any decision of the [Artificial System]." At the same time, even if we considered the case where the person decided to create the artificial system not as a tool, but rather as an end in itself, such a choice would be the person's, and therefore he would be responsible for it. Since the fundamental decision to create the machine as a being was the human's, its freedom would intrinsically be denied, and with that its status as a moral being; therefore the conclusion can be drawn that the very notion of an artificial moral agent represents a *contradictio in adjecto*.

Although employing a different (Kantian influenced) terminology, Bertolini's point is substantially similar to the one offered by Miller: Artifacts are tools that are created for a specific purpose. That purpose cannot be undone by any decision of the artifact nor can it be circumvented by design. Even in those cases where an artifact would be purposefully designed to be something other than a tool, i.e., an object with a purpose, it still has a purpose that was decided by its human designer/maker. For this reason, as Bertolini concludes, "artificial moral agent" is a contradiction in terms. All artifacts are and will remain tools. Even if it is intentionally designed to be something other than a tool, an artifact is still a tool.

6. For this reason, the way that "The Principles" addresses and dispenses with companion robots is potentially problematic. "Consistent with the view of robots as tools, the discussion of robot companionship in the principles is pretty dismissive, describing them as toys that could afford some pleasure to people who are unable, or cannot afford, to keep animal pets. Robots are faux companions on this account that create an 'illusion of emotions' and their intelligence is artificial and not 'real.' The faux nature of robot companions, it is argued, creates a real ethical problem in that robot companions are potentially deceptive and so should be designed so that their 'machine nature is transparent'" (Prescott 2017, 143).

## Chapter 3

1. Like the April 2007 Dana Centre event and preceding press conference at the Science Media Centre (see chapter 1), this BioCenter (http://www.bioethics.ac.uk/) symposium was also developed in direct response to the Ipsos MORI document that

was commissioned and published by the UK Office of Science and Innovation's Horizon Scanning Centre.

2. One can imagine the strange configurations that might need to be presented on an application for approval from IRB (the Internal Review Board, which is charged with evaluating research projects for ethical compliance at US universities, hospitals, and laboratories). Consider, for instance, this hypothetical rationale: "In this experiment, we propose to inflict various levels of pain on an entity in order to prove that it has the capability to feel pain and should, for that reason, have the right to be protected from being subjected to such treatment."

## Chapter 4

1. Bryson (2015) has recognized the problematic connotations that necessarily accompany the use of the word "slave" and has, because of blowback from critics, avoided using it in subsequent work: "I realise now that you cannot use the term 'slave' without invoking its human history, so I don't use this line in talks anymore."

2. One important difference between the argument that is provided by Bryson and the one developed by Miller is how each researcher responds to the capabilities question. For Miller, the question "Can robots have rights?" is a matter of future developments in technology and engineering practices. At some point, he argues, it will be possible to create robots that have humanlike capabilities and at that time such an artifact could be a candidate for rights. For Bryson, the question is more immediate and has to do not with future technological innovations but with legal decisions here and now. According to her argument, robots can have rights to the extent that we can simply decide to define them as moral subjects possessing rights. The difference between these two distinct ways of answering the capabilities question proceeds from, as Bryon points out in the interview with John Danaher, a focus on two different artifacts—the artifact that is the robot and the artifact that is our moral/legal system. This issue will be taken up and further investigated in this chapter.

3. In this context, I use the term "asceticism" in a descriptive and not normative sense. The word "asceticism" is derived from the Greek: ἄσκησις [askesis] denoting "exercise" or "training." "The early Christians adopted it to signify the practice of the spiritual things, or spiritual exercises performed for the purpose of acquiring the habits of virtue" (Campbell 1907, 767). Although the term is often used in an opprobrious sense, it simply describes a manner of intentionally denying oneself something for the sake of attaining a more important objective. Consequently, identifying these approaches as "ascetic" is not meant, in and of itself, to be a criticism or judgment. What is important for the investigation that follows and what is the critical target of its analysis are the specific prohibitions that come to be issued and deployed. In other words, "asceticism" is mobilized here as a description of the kind

of moral procedure that is developed in the texts authored by Bryson, Yampolskiy, and others; it is not utilized as or meant to be a normative judgment.

4. A similar Western/Eastern contrast is developed by Heather Knight (2014, 8), who locates the difference in both religious and literary traditions:

One explanation for the differing cultural response could be religious in origin. The roots of the Western Terminator complex may actually come from the predominance of monotheistic faiths in the West. If it is God's role to create humans, humans that create manlike machines could be seen as usurping the role of God, an act presumed to evince bad consequences. Regardless of current religious practice, such storytelling can permeate cultural expectations. We see this construct in Mary Shelley's story of Frankenstein, first published in 1818. A fictional scientist, Dr. Frankenstein, sews corpses together then brings his super-creature to life in a lightning storm. Upon seeing it animated, he is horrified at the result, and in its abandonment by its creator, the creature turns to ill behavior. This sense of inevitability is cultural, not logical. In Japan, by way of contrast, the early religious history is based on Shintoism. In Shinto animism, objects, animals, and people all share common "spirits," which naturally want to be in harmony. Thus, there is no hierarchy of the species, and left to chance, the expectation is that the outcome of new technologies will complement human society. In the ever-popular Japanese cartoon series Astroboy, we find a very similar formation story to Frankenstein, but the story's cultural environment breeds an opposite conclusion. Astroboy is a robot created by a fictional Ministry of Science to replace the director's deceased son. Initially rejected by that parent figure, he joins a circus where he is rediscovered years later and becomes a superhero, saving society from human flaws.

5. The Japanese perspective, although providing a useful critique of the arguably dominant European influenced account, is not without its own problems and need for criticism. The fact that it is both possible and socially acceptable to grant *koseki* to a robot like Paro while simultaneously withholding the same from a *"zainichi* Korean man who was born, raised, and lives in Japan, who is married to a Japanese citizen, and whose natal family has lived in Japan for generations," indicates a significant challenge for the way that Japanese culture responds to others and other forms of otherness.

6. Please note and let me be absolutely clear on this subject matter. The claim that is made here concerns the texts written and published by Bryson, Miller, Yampolskiy, and EPSRC. It is the text (and specifically the mode of argument that is developed in the text) and not the individual or the community (in the case of EPSRC) that is ethnocentric. In other words, I am NOT saying that Bryson, Miller, Yampolskiy, or EPSRC are ethnocentric. In the case of Miller, Yampolskiy, and most of the individuals in involved in EPSRC, I simply have no idea about or familiarity with their personal opinions, experiences, and positions. With Joanna Bryson, whom I do know personally and have worked with for a number of years, I know for a fact that she is committed to the efforts and objectives of social equality for all human beings. But the argument is not about the individual authors/researchers and what they do or do not believe. That is immaterial in this particular context. The target of the critique that is developed here and throughout the entire book is not the person—this is no *argumentum ad hominem*—but the work that has been produced and published. It is, therefore, NOT about the person; it is about the product.

7. Although this is the concept of slavery that is operationalized in Bryson's text, she does not make explicit reference to Aristotle. This is not necessarily an oversight on her part but more of an indication of the acceptance and "normalization" of the Aristotelian conceptualization. Although Bryson does not take the time (or, perhaps stated more accurately, just did not have the space in the course of a short chapter) to present a detailed consideration of the term "slave," she does offer the following characterization: "*Slaves* are normally defined to be *people you own*" (Bryson 2010, 64; italic in the original). It is a brief comment, but it is one that opens up a number of different (even potentially contradictory) possibilities for how one reads the text. On the one hand, this characterization of "slave" could have the effect of producing an outcome that repudiates the title and thesis of the chapter. Here is how that would go:

1. "*Slaves* are normally defined to be *people you own*" (Bryson 2010, 64).
2. "Robots should not be described as persons, nor given legal or moral responsibility for their actions" (Bryson 2010, 63).
3. Therefore, robots should not be described as slaves.

Formulated in this way, the text could be read as supporting a point that was made by Kathleen Richardson (2016a, 50) in opposition to both Bryson and Aristotle: "It is important to understand that slaves are not machines. Bryson has emphasized that machines are appliances, objects, and these are different from humans, but she draws on the language of slavery, and therefore reproduces Aristotle's assumptions. The confusion about [a] slave's nature stems from Aristotle's misunderstanding of a slave as an animate tool, because slaves were never animate tools." But this interpretation of the text, which would have the effect of asserting the exact opposite of what it sets out to demonstrate (that is, "robots should be slaves"), misses a crucial aspect of Bryson's argument—the distinction between *can* and *should*. It is on the basis of upholding the Humean distinction, even if it is not explicitly recognized as such, that Bryson's essay escapes this alternative reading and the potential charge of self-contradiction: If slaves are people you own, and robots could be defined as persons, then they should be slaves (people you own) and not full (human) persons. Or as Bryson (2010, 65) formulates it: "There [SIC] are the fundamental claims in this paper: (1) Having servants is good and useful, provided no one is dehumanized. (2) A robot can be a servant without being a person. (3) It is right and natural for people to own robots. (4) It would be wrong to let people think that their robots are persons." One of the challenges with following the argument is the fact that there is some equivocation with terminology. "Slave" is both defined as a "person" and characterized as a "servant" that can either be a person or not.

8. The main reason for hesitation comes from the historic fact that slavery, especially within the US context but also across the European "New World" of North and South America, is inextricably tied up with race. There are racial features involved in the term "slave" that cannot be simply set aside, forgotten, or whitewashed. For this reason, reusing or repurposing the term slave as if it could be segre-

gated from this history, and without at least acknowledging the complicated racial dimensions already wired into the concept, risks being insensitive to and complicit with a profound social problem that continues to influence and have an impact on the real lives of individuals and communities in the twenty-first century. Efforts to address and critique the not-so-hidden racial dimensions of robot slavery, although scarcely mentioned in the work of robotics and robot law and ethics, have been increasingly important and visible in science fiction studies: Isiah Lavender's *Race in American Science Fiction* (2011); Gregory Jerome Hampton's *Imagining Slaves and Robots in Literature, Film, and Popular Culture: Reinventing Yesterday's Slave with Tomorrow's Robot* (2015); and Louis Chude-Sokei's *The Sound of Culture: Diaspora and Black Technopoetics* (2015). Additionally, because the institutionalization and experience of African slavery varied significantly across the different European colonies and break-away nation states, one would also need to attend to the way the slave metaphor has been developed and deployed in the context of different cultures and languages—see M. Elizabeth Ginway's *Brazilian Science Fiction: Cultural Myths and Nationhood in the Land of the Future* (2011) and Edward King's *Science Fiction and Digital Technologies in Argentine and Brazilian Culture* (2013).

9. Sharkey et al. (2017, 2) echo this insight: "A problem with the public perception of sex robots is that the public is currently not well informed about the actuality of robots in general. Sex robots are new and only a few people have encountered them directly. Information in the public domain mainly comes from science fiction tropes engendered by television and the movies." Further evidence of the necessary and, for now at least, seemingly unavoidable influence of science fiction is available with Jason Lee's *Sex Robots: The Future of Desire* (2017), which mainly utilizes examples drawn from fiction.

10. If, in fact, the perceived issue was limited to the user of the robot sex-object, the problem could be explained in terms of "individual choice" as is often done with other potentially risky behaviors like smoking and recreational drug use.

11. Whether Richardson herself would characterize this as "slavery" is a complicated question that turns on some inconsistencies in how the term has been defined and utilized in her work. On the one hand, Richardson (2016b, 28) has made a direct connection between sex robots and the Aristotelian understanding of slavery: "For sex robots to become viable possibilities in our culture, it must show us that an Aristotelian sense of the slave must still persist." Although articulated in terms of a conditional statement in this context, a more direct formulation has been reported elsewhere: "An anthropologist and robot ethicist, Richardson claims that owning a sex robot is comparable to owning a slave: individuals will be able to buy the right to only care about themselves, human empathy will be eroded, and female bodies will be further objectified and commodified" (Kleeman 2017). On the other hand, Richardson's article "Robot Sex Matters" (2016a), which begins with a consideration of the price of human bondage, draws a distinction between human slaves and machines, and then criticizes Bryson for conflating the two: "It is important to

understand that slaves are not machines. Bryson has emphasized that machines are appliances, objects, and these are different from humans, but she draws on the language of slavery, and therefore reproduces Aristotle's assumptions" (Richardson 2016a, 50). But in the Ethicomp article that is credited with having launched the "Campaign Against Sex Robots," Richardson appears to endorse the opposite position and incorrectly attributes the opposing view to Bryson: "The use of robots for sex (adults and children) is justified on the basis that robots are not real entities, they are things. ... Is it possible to transfer human constructs of gender, class, race or sexuality to a robot or nonhuman? Anthropologically speaking the answer is yes. This theme has been replaced in a discussion of robots as slaves. Bryson has railed against arguments associating robots with slaves because, she argues, they are nothing more than mechanical appliances—do to robots what you wish. But is it only possible to have an either or position?" (Richardson 2015, 292).

## Chapter 5

1. On the privileged status of the toaster as the example of examples, see chapter 2.

2. Robot abuse studies, like that undertaken by Darling and Gassert during the LIFT13 workshop have been around for a number of years (cf. De Angeli, Brahnam, and Wallis 2005; De Angeli, Brahnam, Wallis, and Dix 2006; Brahnam and De Angeli 2008) and comprise what has been described as "the dark side" of research in HCI (human-computer interaction) and HRI (human-robot interaction) (De Angeli, Brahnam, and Wallis 2005). As Bartnek and Hu (2008) explain, the reason for this kind of work is to examine the limits of human/robot interaction by intentionally stepping outside the boundaries of normal conduct: "It is only from an extreme position that the applicability of the Media Equation to robots might become clear. In our study we have therefore focused on robot abuse. What we propose to investigate in this context is whether human beings abuse robots in the same way as they abuse other human beings, as suggested by the Media Equation" (Bartnek and Hu 2008, 416). Although previous efforts like those of De Angeli, Brahnam, Wallis, and Dix (2006, 1) focus on design implications—"the overarching objective ... is to sketch a research agenda on the topic of the misuse and abuse of interactive technologies that will lead to design solutions capable of protecting users and restraining disinhibited behaviors"—Darling redeploys the approach to articulate and address questions of moral and legal standing.

3. This normative component is, in fact, one of the distinguishing characteristics of Darling's own efforts. Although there have been numerous studies of the media equation with computers and related systems (e.g., Reeves and Nass 1996, Nass et al. 1997, Nass and Moon 2000), laboratory investigations of the opportunities and challenges of HRI (e.g., Eyssel and Kuchenbrandt 2012, Kahn et al. 2012, Faragó et al. 2014, Broadbent 2017), and Milgram-like obedience experiments involving robot abuse (e.g., Bartnek and Hu 2008 and Rosenthal-von der Pütten et al. 2013), these

investigations have typically stopped short of drawing any explicit normative conclusions about, or offering any advice concerning, the social position and legal status of robots. This is, however, what interests Darling and constitutes her unique contribution to the field. To put it another way, the numerous social scientific investigations conducted over the past several decades demonstrate that human users do in fact respond to and treat robots in ways that are demonstratively differently from how they respond to other kinds of objects, like a toaster. Darling simply takes up these descriptive insights and pursues their normative consequences.

4. In her previous book, *Life on the Screen*, Turkle (1995, 321) does provide a short reflection on this particular method of investigation: "This is a very personal book. It is based on ethnographic and clinical observations, where the researcher, her sensibilities and tastes, constitute a principal instrument of investigation. My method is drawn from the various formal disciplines I have studied and grown up in; the work itself is motivated by a desire to convey what I have found most compelling and significant in the evolving relationship between computers and people." Because "the researcher, her sensibilities and tastes" comprise a "principal instrument of investigation," Turkle's "A Note on Method" (a chapter in *Life on the Screen*) provides a rather detailed account of the researcher's own involvement with computers and information technology. Although this personal history is interesting and informative, the note identifies and confirms the problem that is cited by Jones, namely that Turkle's insights remain difficult to generalize insofar as they are based on and limited to personal experiences and individual sensibilities.

5. This is precisely the organizing principle and *raison d'etre* of the popular television series *Robot Wars* (1998–2004 and 2016–present), which stages violent conflicts between robots (and, by extension, the teams of human beings who designed and built the robot) for the sake of (human) entertainment. Whether it is in fact the case that "observing robots abusing each other could lead to psychological trauma" (Ashrafian 2015a, 36) or not, remains an open question.

6. Or as Darling (2016a, 224) characterizes it: "we may soon have to consider whether or not to permit sexual practices between humans and social robots that we currently do not permit when the receiver is a live human or animal. Bestiality, rape, and in particular sexual acts with underage children are condemned in our culture and heavily governed by our legal system. It is thinkable that the desire to protect our current social values could cause people to demand laws that prohibit the sexual abuse of social robots."

7. Although not necessarily offered in support of the position, here is how Danaher (2017a, 90) describes the logic involved in this way of thinking: "Since we are assuming there is no moral victim in the robotic case, real-world rape and child sexual abuse are obviously much worse than their robotic equivalents. Any reduction in the risk of real-world acts would be worth it. I would suggest that the possibility of such a reduction should be actively and carefully researched. It would have

important benefits for society, and also for those in the grip of certain forms of paraphilia. People already advocate much more drastic therapies for such people—e.g. chemical castration. The permitted use of a sex robot would be less invasive, and arguably preferable."

8. In an effort to contend with this undecidability, Danaher (2017a, 95) not only recommends that any "extrinsic argument" needs to be extremely careful with "obtaining good evidence relating to the extrinsic effects of using such sex robots" but also proposes a "plausible *prima facie* argument" for the criminalization of robot rape and child abuse that does not utilize or depend upon these extrinsic conditions and consequences.

## Chapter 6

1. On the significance of the phrase "thinking otherwise" and the (non)method of deconstruction, see the Introduction.

2. Although it is beyond the scope of this analysis, it would be worth comparing the philosophical innovations of Emmanuel Levinas to the efforts of Knud Ejler Løgstrup. In their "Introduction" to the English translation of Løgstrup's *The Ethical Demand*, Alasdair MacIntyre and Hans Fink offer the following comment:

Zygmunt Bauman in his *Postmodern Ethics* (Oxford: Blackwell, 1984)—a remarkable conspectus of a number of related postmodern standpoints—presents Løgstrup's work as having close affinities with Emmanuel Levinas. (Levinas was teaching at Strasbourg when Løgstrup was there in 1930, but there is no evidence that Løgstrup attended his Lectures.) Levinas, who was one of Husserl's students, was from the outset and has remained much closer to Husserlian phenomenology than Løgstrup ever was, often defining his own positions, even when they are antagonistic to Husserl's, in terms of their relationships to Husserl's. But, on some crucial issues, as Bauman's exposition makes clear, Levinas and Løgstrup are close. (MacIntyre and Fink 1997, xxxiii)

MacIntyre and Fink continue the comparison by noting important similarities that occur in the work of both Levinas and Løgstrup concerning the response to the Other, stressing that responsibility "is not derivable from or founded upon any universal rule or set of rights or determinate conception of the human good. For it is more fundamental in the moral life than any of these" (MacIntyre and Fink 1997, xxxiv).

3. Although this might look like "name dropping," and in one sense it is, the individuals Taylor references in this quotation are the widely recognizable "thought leaders" in poststructuralism. Jacques Derrida (1930–2004), Jacques Lacan (1901–1981) and Michel Foucault (1926–1984) are listed here as representatives of what might be called the "philosophical wing" of poststructuralism, while Hélèn Cixous (1937–), Julia Kristeva (1941–), and Michel de Certeau (1925–1986) represent the "social sciences" side, coming from the disciplines of literary theory, semiology, and sociology, respectively. Although not listed by name, Levinas is commonly associated with the first group.

4. For a critical reading and evaluation of IE and its ontocentric formulations, cf. Gunkel (2012).

5. "The concept of responsibility," as Paul Ricœur (2007, 11) insightfully pointed out in his eponymously titled essay, is anything but clear and well defined. Although the classical juridical usage of the term, which dates back to the nineteenth century, seems rather well-established—with "responsibility" characterized in terms of both civil and penal obligations (either the obligation to compensate for harms or the obligation to submit to punishment)—the philosophical concept remains confused and somewhat vague.

> In the first place, we are surprised that a term with such a firm sense on the juridical plane should be of such recent origin and not really well established within the philosophical tradition. Next, the current proliferation and dispersion of uses of this term is puzzling, especially because they go well beyond the limits established for its juridical use. The adjective "responsible" can complement a wide variety of things: you are responsible for the consequences of your acts, but also responsible for others' actions to the extent that they were done under your charge or care ... In these diffuse uses the reference to obligation has not disappeared, it has become the obligation to fulfill certain duties, to assume certain burdens, to carry out certain commitments. (Ricœur 2007, 11–12)

Ricœur (2007, 12) traces this sense of the word through its etymology (hence the subtitle to the essay "A Semantic Analysis") to "the polysemia of the verb 'to respond'," which denotes "to answer for ..." or "to respond to ... (a question, an appeal, an injunction, etc.)." It is this sense of the word, and not the more restricted juridical usage, that is operative and operationalized in Levinasian philosophy.

6. A similar remark has been offered by Anthony Beavers (1995, 109): "We must remember, however, that Levinas never actually presents an ethics. In fact, he writes, 'One can without doubt construct an ethics in function of what I have ... said, but this is not my own theme' (Levinas 1985, 90). ... His [Levinas's] project located the origin of moral responsibility; he determined how freedom becomes endowed with its ought. He did not determine what this means for normative considerations, and admittedly, it is difficult at the outset to see precisely how one might get from Levinas' metaphysical grounding of moral responsibility to an ethics." But unlike Derrida and others, who take note of this fact and pursue its consequences, Beavers endeavors to construct a normative ethics from Levinas's "proto-ethics." As Beavers (1995, 109) explains: "We must begin to complete Levinas' account of the self in relation to the other person in order to establish ethical reciprocity. Once the self is viewed as the other's other—that is, as an other person who issues a moral command to the other person—the possibility of mutual relationship opens up, and with it, the possibility of ethical norms to balance the needs and desires of the self against those of the other." Although this might sound like a promising proposal, these standard metaphysical concepts, like "reciprocity" and "mutual relationship," are inherently abrasive to and dissonant with Levinasian philosophy insofar as Levinas has insisted, from the very beginning, that the ethical relationship—the eruption of and encounter with the face of the other—is and remains fundamentally

asymmetrical. "Levinas teaches," Duncan (2006, 271) explains, "that ethical responsibility goes from me to the neighbor and is not part of a symmetrical determination in which I could say that an equal responsibility weighs on both of us, or insofar as I can ascribe responsibility to my neighbor, I find myself responsible for his responsibility. There is no evening up the score."

7. For a critical examination of the "relational-turn" applied to the problematic of animal rights philosophy, see Coeckelbergh and Gunkel "Facing Animals: A Relational, Other-Oriented Approach to Moral Standing" (2014), the critical commentary provided by Michał Piekarski (2016), and Coeckelbergh and Gunkel's (2016) reply to Piekarski's criticisms.

8. This colloquialism, although useful in this particular context, is something that is also subjected to Levinas's critical engagement with the history of Western philosophy. The structuring principle of θεωρία [*theoria*], which is always a matter of looking and framing how one looks at things, the Platonic tradition of "the light of being," and the legacy and logic of the enlightenment are all part of an ideology that is targeted and problematized in Levinas's work.

9. As of this writing (November 2017), units of Jibo have finally begun shipping. Despite initial optimism concerning the production schedule and an impressive number of preorders (partially due to the success of the promotional video), there had been considerable delays in the delivery of the product.

10. This is one of the points of contact and contention between Levinas and Martin Buber. For Levinas, Buber's *ich/du* [I/You] relationship already risks conceptualizing Other as alter ego, thereby violating and doing violence to the alterity of others. For a thorough investigation of the rather intricate debate between Levinas and Buber, see Peter Atterton, Matthew Calarco, and Maurice Friedman's edited collection *Levinas and Buber: Dialogue and Difference* (2004).

11. "Difficult to think," it should be recalled, not only because such thinking is "hard," insofar as it challenges the very conceptuality of thought itself, but also because it is, for many thinkers, hard to keep a "straight face" (see chapter 1) in the face of this unthinkable, if not "ridiculous," thought.

# References

Abney, Keith. 2012. Robotics, ethical theory, and metaethics: A Guide to the perplexed. In *Robot Ethics: The Ethical and Social Implications of Robotics*, ed. Patrick Lin, Keith Abney, and George A. Bekey, 35–52. Cambridge, MA: MIT Press.

Adams, Brian, Cynthia L. Breazeal, Rodney Brooks, and Brian Scassellati. 2000. Humanoid robots: A new kind of tool. *IEEE Intelligent Systems* 15 (4): 25–31. doi:10.1109/5254.867909.

Anderson, Susan Leigh. 2008. Asimov's "three laws of robotics" and machine metaethics. *AI & Society* 22 (4): 477–493. doi:10.1007/s00146-007-0094-5.

Anderson, Michael, and Susan Leigh Anderson. 2007. Machine ethics: Creating an ethical intelligent agent. *AI Magazine* 28 (4): 15–26. doi:10.1609/aimag.v28i4.2065.

Anderson, Michael, and Susan Leigh Anderson. 2008. EthEl: Toward a principled ethical eldercare robot. In *Proceedings of the AAAI Fall 2008 Symposium on AI in Eldercare: New Solutions to Old Problems*, 4–11. Menlo Park, CA: AAAI Press.

Anderson, Michael, and Susan Leigh Anderson, eds. 2011. *Machine Ethics*. Cambridge: Cambridge University Press.

Anderson, Michael, Susan Leigh Anderson, and Chris Armen. 2004. Toward machine ethics. *American Association for Artificial Intelligence*. http://www.aaai.org/Papers/Workshops/2004/WS-04-02/WS04-02-008.pdf.

Aristotle. 1944. *Politics*. Trans. H. Rackham. Cambridge, MA: Harvard University Press.

Arkin, Ronald. 2009. *Governing Lethal Behavior in Autonomous Robots*. Boca Raton, FL: Chapman and Hall/CRC.

Asada, Minoru, Karl F. MacDorman, Hiroshi Ishiguro, and Yasuo Kuniyoshi. 2001. Cognitive developmental robotics as a new paradigm for the design of humanoid robots. *Robotics and Autonomous Systems* 37 (2–3): 185–193. doi:10.1016/S0921-8890(01)00157-9.

Asaro, Peter. 2006. What should we want from a robot ethic? *International Review of Information Ethics* 6 (12): 9–16. http://www.i-r-i-e.net/inhalt/006/006_full.pdf.

Asaro, Peter. 2012. A body to kick, but still no soul to damn: Legal perspectives on robotics. In *Robot Ethics: The Ethical and Social Implications of Robotics*, ed. Patrick Lin, Keith Abney and George A. Bekey, 169–186. Cambridge, MA: MIT Press.

Asaro, Peter, and Wendell Wallach. 2016. Introduction: The emergence of robot ethics and machine ethics. In *Machine Ethics and Robot Ethics*, ed. Peter Asaro and Wendell Wallach, 1–15. New York: Routledge.

Ashrafian, Hutan. 2015a. AIonAI: A humanitarian law of artificial intelligence and robotics. *Science and Engineering Ethics* 21 (1): 29–40. doi:10.1007/s11948-013-9513-9.

Ashrafian, Hutan. 2015b. Artificial intelligence and robot responsibilities: Innovating beyond rights. *Science and Engineering Ethics* 21 (2): 317–326. doi:10.1007/s11948-014-9541-0.

Asimov, Isaac. 1981. *Asimov on Science Fiction*. Garden City, NY: Doubleday.

Atterton, Peter, Matthew Calarco, and Maurice Friedman, eds. 2004. *Levinas and Buber: Dialogue and Difference*. Pittsburgh, PA: Duquesne University Press.

Atterton, Peter, and Matthew Calarco, eds. 2010. *Radicalizing Levinas*. Albany, NY: SUNY Press.

Bartneck, Christoph. 2004. From fiction to science—A cultural reflection of social robots. *Proceedings of the CHI 2004 Workshop on Shaping Human-Robot Interaction*, Vienna.

Bartnek, Christoph, and Jun Hu. 2008. Exploring the abuse of robots. *Interaction Studies: Social Behaviour and Communication in Biological and Artificial Systems* 9 (3): 415–433. doi:10.1075/is.9.3.04bar.

Bauman, Zygmunt. 1984. *Postmodern Ethics*. Oxford: Blackwell.

BBC News. 2006. Robots could demand legal rights. *BBC News*, December 21. http://news.bbc.co.uk/2/hi/technology/6200005.stm.

Beavers, Anthony F. 1995. *Levinas Beyond the Horizons of Cartesianism: An Inquiry into the Metaphysics of Morals*. New York: Peter Lang.

Beavers, Anthony F. 2012. Moral machines and the threat of ethical nihilism. In *Robot Ethics: The Ethical and Social Implications of Robotics*, ed. Patrick Lin, Keith Abney and George A. Bekey, 333–344. Cambridge, MA: MIT Press.

Bekey, George A. 2015. *Autonomous Robots: From Biological Inspiration to Implementation and Control*. Cambridge, MA: MIT Press.

Bekoff, Marc. 2007. *The Emotional Lives of Animals: A Leading Scientist Explores Animal Joy, Sorrow, and Empathy—And Why They Matter*. Novato, CA: New World Library.

Bemelmans, Roger, Gert Jan Gelderblom, Pieter Jonker, and Luc de Witte. 2010. The potential of socially assistive robotics in care for elderly, a systematic review. In *Human-Robot Personal Relationships*, ed. Maarten H. Lamers and Fons J. Verbeek, 83–89. Heidelberg: Springer.

Benford, Gregory, and Elisabeth Malartre. 2007. *Beyond Human: Living with Robots and Cyborgs*. New York: Tom Doherty.

Benso, Silvia. 2000. *The Face of Things: A Different Side of Ethics*. Albany, NY: SUNY Press.

Bensoussan, Alain. 2014. Le droit des robots: Mythe ou réalité? *Emile & Ferdinand* (October). https://www.alain-bensoussan.com/wp-content/uploads/2014/11/28743700.pdf.

Bensoussan, Alain and Jérémy Bensoussan. 2015. *Droit des robots*. [The Law of Robots.] Brussels: Éditions Larcier.

Bensoussan, Alain and Jérémy Bensoussan. 2016. *Comparative Handbook: Robotic Technologies Law*. Brussels: Larcier Group.

Bentham, Jeremy. 2005. *An Introduction to the Principles of Morals and Legislation*. Ed. J. H. Burns and H. L. Hart. Oxford: Oxford University Press.

Beran, Tanya N., and Alejandro Ramirez-Serrano. 2010. Can children have a relationship with a robot? In *Human-Robot Personal Relationships*, ed. Maarten H. Lamers and Fons J. Verbeek, 49–56. Heidelberg: Springer.

Bertolini, Andrea. 2013. Robots as products: The case for a realistic analysis of robotic applications and liability rules. *Law Innovation and Technology* 5 (2): 214–247. doi:10.5235/17579961.5.2.214.

Bhuta, Nehal, Susanne Beck, Robin Geiß, Hin-Yan Liu, and Claus Kreß. 2016. *Autonomous Weapons Systems: Law, Ethics, Policy*. Cambridge: Cambridge University Press.

Binder, O. O. 1957. You will own "slaves" by 1965. *Mechanix Illustrated* (January): 62–65.

Birch, Thomas. 1993. Moral considerability and universal consideration. *Environmental Ethics* 15:313–332. doi:10.5840/enviroethics19931544.

Birch, Thomas H. 1995. The incarnation of wilderness: Wilderness areas as prisons. In *Postmodern Environmental Ethics*, ed. Max Oelschlaeger, 137–162. Albany, NY: SUNY Press.

Bishop, Mark. 2009. Why computers can't feel pain. *Minds and Machines* 19 (4): 507–516. doi:10.1007/s11023-009-9173-3.

Bobbio, Norberto. 1996. *The Age of Rights*. Trans. A. Cameron. Cambridge: Polity Press.

Boden, Margaret, Joanna Bryson, Darwin Caldwell, Kerstin Dautenhahn, Lilian Edwards, Sarah Kember, Paul Newman, et al. 2011. Principles of robotics: Regulating robots in the real World. *Engineering and Physical Sciences Research Council* (EPSRC). https://www.epsrc.ac.uk/research/ourportfolio/themes/engineering/activities/principlesofrobotics.

Boden, Margaret, Joanna Bryson, Darwin Caldwell, Kerstin Dautenhahn, Lilian Edwards, Sarah Kember, Paul Newman, et al. 2017. Principles of robotics: Regulating robots in the real world. *Connection Science* 29 (2): 124–129. doi:10.1080/09540091.2016.1271400.

Bostrom, Nick. 2014. *Superintelligence: Paths, Dangers, Strategies*. New York: Oxford University Press.

Bostrom, Nick, and Eliezer Yudkowsky. 2014. The ethics of artificial intelligence. In *The Cambridge Handbook of Artificial Intelligence*, ed. Keith Frankish and William M. Ramsey, 316–334. Cambridge: Cambridge University Press.

Brahnam, Sheryl, and Antonella De Angeli. 2008. Special issue: Abuse and misuse of social agents. *Interacting with Computers* 20 (3): 287–291. doi:10.1016/j.intcom.2008.02.001.

Breazeal, Cynthia L. 2004. *Designing Sociable Robots*. Cambridge, MA: MIT Press.

Broadbent, Elizabeth, Vinayak Kumar, Xingyan Li, John Sollers, Rebecca Q. Stafford, Bruce A. MacDonald, and Daniel M. Wegner. 2013. Robots with display screens: A robot with a more humanlike face display is perceived to have more mind and a better personality. *PLoS One* 8 (8): e72589.

Brooks, Rodney A. 2002. *Flesh and Machines: How Robots Will Change Us*. New York: Pantheon Books.

Bruckenberger, Ulrike, Astrid Weiss, Nicole Mirnig, Ewald Strasser, Susanne Stadler, and Manfred Tscheligi. 2013. The good, the bad, the weird: Audience evaluation of a "real" robot in relation to science fiction and mass media. In *Social Robotics. ICSR 2013*, vol. 8239. Ed. Guido Herrmann, Martin J. Pearson, Alexander Lenz, Paul Bremner, Adam Spiers, and Ute Leonards, 301–310. Lecture Notes in Computer Science. Cham, Switzerland: Springer. doi:10.1007/978-3-319-02675-6_30.

Brynjolfsson, Erik, and Andrew McAfee. 2011. *Race Against the Machine: How the Digital Revolution is Accelerating Innovation, Driving Productivity, and Irreversibly Transforming Employment and the Economy*. Lexington, MA: Digital Frontier Press.

Bryson, Joanna. 2000. A proposal for the humanoid agent-builder's league (HAL). In *Proceedings of The AISB 2000 Symposium on Artificial Intelligence, Ethics and (Quasi-)*

*Human Rights*, ed. John Barnden and Mark Lee, 1–6. Hove, UK: AISB. http://www.aisb.org.uk/publications/proceedings/aisb2000/AISB00_Ethics.pdf.

Bryson, Joanna. 2010. Robots should be slaves. In *Close Engagements with Artificial Companions: Key Social, Psychological, Ethical and Design Issues*, ed. Yorick Wilks, 63–74. Amsterdam: John Benjamins.

Bryson, Joanna. 2012. A role for consciousness in action selection. *International Journal of Machine Consciousness* 4 (2): 471–482. doi:10.1142/S1793843012400276.

Bryson, Joanna. 2015. Clones should NOT be slaves. *Adventures in NI* (4 October). https://joanna-bryson.blogspot.com/2015/10/clones-should-not-be-slaves.html.

Bryson, Joanna. 2016a. Patiency is not a virtue: AI and the design of ethical systems. AAAI Spring Symposium Series. Ethical and Moral Considerations in Non-Human Agents. http://www.aaai.org/ocs/index.php/SSS/SSS16/paper/view/12686.

Bryson, Joanna. 2016b. Robots are owned. Owners are taxed. Internet services cost Information. *Adventures in NI* (June 23). https://joanna-bryson.blogspot.com/2016/06/robots-are-owned-owners-are-taxed.html.

Bryson, Joanna, and Philip P. Kime. 2011. Just an artifact: Why machines are perceived as moral agents. In the *Proceedings of The Twenty-Second International Joint Conference on Artificial Intelligence*. Barcelona, Spain. https://www.cs.bath.ac.uk/~jjb/ftp/BrysonKime-IJCAI11.pdf.

Bryson, Joanna J., Mihailis E. Diamantis, and Thomas D. Grant. 2017. Of, for, and by the people: The legal lacuna of synthetic persons. *Artificial Intelligence and Law* 25 (3): 273–291. doi:10.1007/s10506-017-9214-9.

Calarco, Matthew. 2008. *Zoographies: The Question of the Animal from Heidegger to Derrida*. New York: Columbia University Press.

Callicott, J. Baird. 1989. *In Defense of the Land Ethic: Essays in Environmental Philosophy*. Albany: State University of New York Press.

Calo, Ryan. 2011. The sorcerer's apprentice or: Why weak AI is interesting enough. *The Center for Internet and Society*. August 11. http://cyberlaw.stanford.edu/blog/2011/08/sorcerers-apprentice-or-why-weak-ai-interesting-enough.

Calo, Ryan, A. Michael Froomkin, and Ian Kerr, eds. 2016. *Robot Law*. Cheltenham, UK: Edward Elgar Publishing.

Calverley, D. 2005. Toward a method for determining the legal status of a conscious machine. In *Proceedings of the AISB 2005 symposium on next generation approaches to machine consciousness: Imagination, development, intersubjectivity, and embodiment*, ed. R. Chrisley, R. Clowes and S. Torrance. Hatfield, UK: University of Hertfordshire. http://www.aisb.org.uk/publications/proceedings/aisb2005/7_MachConsc_Final.pdf.

Campaign Against Sex Robots. 2016. https://campaignagainstsexrobots.org.

Campbell, T. J. 1907. Asceticism. In *The Catholic Encyclopedia: An International Work of Reference on the Constitution, Doctrine, Discipline, and History of the Catholic Church*. Vol. I. Ed. Charles G. Herbermann, Edward A. Pace, Conde B. Pallen, Thomas J. Shahan, and John J. Wynne, 767–773. New York: Robert Appleton Company.

Campbell, Tom. 2006. *Rights: A Critical Introduction*. New York: Routledge.

Camus, Albert. 1983. *The Myth of Sisyphus, and Other Essays*. Trans. J. O'Brien. New York: Alfred A. Knopf.

Čapek, Karel. 2009. *R.U.R. (Rossum's Universal Robots)*. Trans. D. Wyllie. Gloucestershire, UK: The Echo Library.

Čapek, Karel. 1935. The author of robots defends himself—Karel Čapek. *Lidove Noviny* 43 (290), June 9. Collected in Karel Čapek. 1986. *O Umení a Kulture III* (Spisy XIX), 656–657. Trans. by Cyril Simsa. Prague: Ceskoslovensky spisovatel. http://www.depauw.edu/sfs/documents/capek68.htm.

Carpenter, Julie. 2015. *Culture and Human-Robot Interaction in Militarized Spaces: A War Story*. New York: Ashgate.

Casey, Bryan James. 2017. Amoral machines, or: How roboticists can learn to stop worrying and love the law. *Northwestern University Law Review Online* 111: 231–250. http://scholarlycommons.law.northwestern.edu/cgi/viewcontent.cgi?article=1248&context=nulr_online.

Channell, David F. 1991. *The Vital Machine: A Study of Technology and Organic Life*. Oxford: Oxford University Press.

Chella, Antonio, and Riccardo Manzotti. 2009. Machine consciousness: A manifesto for robotics. *International Journal of Machine Consciousness* 1 (1): 33–51. doi:10.1142/S1793843009000062.

Chicago Manual of Style (CMS). 2017. FAQ—Pronouns. http://www.chicagomanualofstyle.org/qanda/data/faq/topics/Pronouns/faq0020.html.

Chu, Seo-Young. 2010. *Do Metaphors Dream of Literal Sleep? A Science-Fictional Theory of Representation*. Cambridge, MA: Harvard University Press.

Chude-Sokei, Louis. 2015. *The Sound of Culture: Diaspora and Black Technopoetics*. Middleton, CT: Wesleyan University Press.

Coeckelbergh, Mark. 2010a. Robot rights? Towards a social-relational justification of moral consideration. *Ethics and Information Technology* 12 (3): 209–221. doi:10.1007/s10676-010-9235-5.

Coeckelbergh, Mark. 2010b. Moral appearances: Emotions, robots, and human morality. *Ethics and Information Technology* 12 (3): 235–241. doi:10.1007/s10676-010-9221-y.

# References

Coeckelbergh, Mark. 2011. You, robot: On the linguistic construction of artificial others. *AI & Society* 26 (1): 61–69. doi:10.1007/s00146-010-0289-z.

Coeckelbergh, Mark. 2012. *Growing Moral Relations: Critique of Moral Status Ascription.* New York: Palgrave MacMillan.

Coeckelbergh, Mark. 2014. Robotic appearance and forms of life: A phenomenological-hermeneutical approach to the relation between robotics and culture. In *Robotics in Germany and Japan: Philosophical and Technical Perspectives*, ed. Michael Funk and Bernhard Irrgang, 59–68. Frankfurt am Main: Peter Lang.

Coeckelbergh, Mark. 2016a. Is it wrong to kick a robot? Towards a relational and critical robot ethics and beyond. In *What Social Robots Can and Should Do*, ed. Johanna Seibt, Marco Nørskov, and Søren Schack Andersen, 7–8. Amsterdam: IOS Press.

Coeckelbergh, Mark. 2016b. Why care about robots? Empathy, moral standing and the language of suffering. Keynote address at *The Third International Conference on Philosophy of Science: Contemporary Issues*. December 14–16. Lisbon, Portugal. http://lisbonicpos.campus.ciencias.ulisboa.pt/keynote-speakers.

Coeckelbergh, Mark. 2016c. Alterity ex machina: The encounter with technology as an epistemological-ethical drama. In *The Changing Face of Alterity: Communication, Technology and Other Subjects*, ed. David J. Gunkel, Ciro Marcondes Filho and Dieter Mersch, 181–196. New York: Rowman & Littlefield.

Coeckelbergh, Mark. 2017. *Using Words and Things: Language and Philosophy of Technology.* New York: Routledge.

Coeckelbergh, Mark, and David J. Gunkel. 2014. Facing animals: A relational, other-oriented approach to moral standing. *Journal of Agricultural & Environmental Ethics* 27 (5): 715–733. doi:10.1007/s10806-013-9486-3.

Coeckelbergh, Mark, and David J. Gunkel. 2016. Response to "the problem of the question about animal ethics" by Michal Piekarski. *Journal of Agricultural & Environmental Ethics* 29 (4): 717–721. doi:10.1007/s10806-016-9627-6.

Cohen, Richard A. 2000. Ethics and cybernetics: Levinasian reflections. *Ethics and Information Technology* 2:27–35. doi:10.1023/A:1010060128998.

Cohen, Richard A. 2003. Introduction. In *Humanism of the Other*, trans. Nidra Poller, vii—xliv. Urbana, IL: University of Illinois Press.

Cohen, Richard A. 2010. Ethics and cybernetics: Levinasian reflections. In *Radicalizing Levinas*, ed. Peter Atterton and Matthew Calarco, 153–170. Albany, NY: SUNY Press.

Collins, Kate. 2015. What the death of HitchBOT teaches us about humans. *Wired.* http://www.wired.co.uk/article/hitchbot-death-robot-ethics-human-psychology.

Committee on Legal Affairs. 2016. Draft report with recommendations to the commission on civil law rules on robotics. European Parliament. http://www.europarl.europa.eu/sides/getDoc.do?pubRef=-//EP//NONSGML%2BCOMPARL%2BPE-582.443%2B01%2BDOC%2BPDF%2BV0//EN.

Cookson, Clive. 2015. Scientists appeal for ethical use of robots. *Financial Times*, December 10. https://www.ft.com/content/fee8bacc-9f37-11e5-8613-08e211ea5317.

Culler, Jonathan. 1982. *On Deconstruction: Theory and Criticism After Structuralism*. Ithaca, NY: Cornell University Press.

Cushman, Fiery, Kurt Gray, Allison Gaffey, and Wendy Berry Mendes. 2012. Simulating murder: The aversion to harmful action. *Emotion (Washington, D.C.)* 12 (1): 2–7.

Danaher, John. 2016. Robots, law and the retribution gap. *Ethics and Information Technology* 18 (4): 299–309. doi:10.1007/s10676-016-9403-3.

Danaher, John. 2017a. Robotic rape and robotic child sexual abuse: Should they be criminalised? *Criminal Law and Philosophy* 11 (1): 71–95. doi:10.1007/s11572-014-9362-x.

Danaher, John. 2017b. Should robots have rights? Four perspectives. *Philosophical Disquisitions*. http://philosophicaldisquisitions.blogspot.com/2017/10/should-robots-have-rights-four.html.

Danaher, John, and Joanna Bryson. 2017. Bryson on why robots should be slaves. *Philosophical Disquisitions*, podcast #24. http://philosophicaldisquisitions.blogspot.com/2017/06/episode-24-bryson-on-why-robots-should.html.

Danaher, John, and Neil McArthur. 2017. *Robot Sex: Social and Ethical Implications*. Cambridge, MA: MIT Press.

Darling, Kate. 2012. Extending legal protection to social robots. *IEEE Spectrum*. http://spectrum.ieee.org/automaton/robotics/artificial-intelligence/extending-legal-protection-to-social-robots.

Darling, Kate. 2015. Ethical issues in human-robot interaction. Media Evolution / The Conference. https://www.youtube.com/watch?v=m3gp4LFgPX0.

Darling, Kate. 2016a. Extending legal protection to social robots: The effects of anthropomorphism, empathy, and violent behavior toward robotic objects. In *Robot Law*, ed. Ryan Calo, A. Michael Froomkin, and Ian Kerr, 213–231. Northampton, MA: Edward Elgar.

Darling, Kate. 2016b. What are the rules of human-robot interaction? SBS Seoul Digital Forum. https://www.youtube.com/watch?v=k7NNK-nQquo.

Darling, Kate. 2017. "Who's Johnny?" Anthropomorphic framing in human-robot interaction, integration, and policy. In *Robot Ethics 2.0: From Autonomous Cars to*

*Artificial Intelligence*, ed. Patrick Lin, Ryan Jenkins, and Keith Abney, 173–191. New York: Oxford University Press.

Darling, Kate, and Sabine Hauert. 2013. Giving rights to robots. *RobotsPodcast #125*. http://robohub.org/robots-giving-rights-to-robots.

Darling, Kate, and Shankar Vedantam. 2017. Can robots teach us what it means to be human? *Hidden Brain* (NPR Podcast). http://www.npr.org/2017/07/10/536424647/can-robots-teach-us-what-it-means-to-be-human.

Darling, Kate, Palash Nandy, and Cynthia Breazeal. 2015. Empathic concern and the effect of stories in human-robot interaction. *Proceedings of the 24th IEEE International Symposium on Robot and Human Interactive Communication (RO-MAN)*, ed. IEEE, 770–775. doi: 10.1109/ROMAN.2015.7333675. Also available at https://papers.ssrn.com/sol3/papers.cfm?abstract_id=2639689.

Dashevsky, Evan. 2017. Do robots and AI deserve rights? *PC Magazine*, February 16. https://www.pcmag.com/article/351719/do-robots-and-ai-deserve-rights.

Dautenhahn, Kirsten. 1998. The art of designing socially intelligent agents: Science, fiction, and the human in the loop. *Applied Artificial Intelligence* 12 (7–8): 573–617. doi:10.1080/088395198117550.

Davoudi, Salamander. 2006. UK report says robots will have rights. *Financial Times*, December 19. https://www.ft.com/content/5ae9b434-8f8e-11db-9ba3-0000779e2340.

Davy, Barbara Jane. 2007. An other face of ethics in Levinas. *Ethics and the Environment* 12 (1): 39–65. http://www.jstor.org/stable/40339131.

De Angeli, Antonella, Sheryl Brahnam, and Peter Wallis. 2005. Abuse: The dark side of human—computer interaction. *Interact* 2005. http://www.agentabuse.org.

De Angeli, Antonella, Sheryl Brahnam, Peter Wallis, and Alan Dix. 2006. Misuse and abuse of interactive technologies. Proceedings of the CHI '06 extended abstracts on human factors in computing systems, 22–27 April 2006, Montreal, Québec, Canada. http://www.agentabuse.org.

Deleuze, Gilles. 1994. *Difference and Repetition*. Trans. P. Patton. New York: Columbia University Press.

Dennett, Daniel. 1978. Why you can't make a computer that feels pain. *Synthese* 38 (3): 415–456. doi:10.1007/BF00486638.

Dennett, Daniel. 1996. *Kinds of Minds*. New York: Basic Books.

Dennett, Daniel C. 1997. When HAL kills, who's to blame? Computer ethics. In *Hal's Legacy: 2001's Computer as Dream and Reality*, ed. David G. Stork, 351–366. Cambridge, MA: MIT Press.

Dennett, Daniel C. 1998. *Brainstorms: Philosophical Essays on Mind and Psychology*. Cambridge, MA: MIT Press.

Dennett Daniel, C. 2001. Are we explaining consciousness yet? *Cognition* 79 (1–2): 221–237. doi:10.1016/S0010-0277(00)00130-X.

Dennett Daniel, C. 2009. The part of cognitive science that is philosophy. *Topics in Cognitive Science* 1 (2): 231–236. doi:10.1111/j.1756-8765.2009.01015.x.

Derrida, Jacques. 1978. *Writing and Difference*. Trans. A. Bass. Chicago: University of Chicago Press.

Derrida, Jacques. 1981. *Positions*. Trans. A. Bass. Chicago: University of Chicago Press.

Derrida, Jacques. 1982. *Margins of Philosophy*. Trans. A. Bass. Chicago: University of Chicago Press.

Derrida, Jacques. 1991. Letter to a Japanese friend. Trans. D. Wood and A. Benjamin. In *A Derrida Reader: Between the Blinds*, ed. Peggy Kamuf, 270–276. New York: Columbia University Press.

Derrida, Jacques. 1993. *Limited Inc*. Trans. S. Weber and J. Mehlman. Evanston, IL: Northwestern University Press.

Derrida, Jacques. 2005. *Paper Machine*. Trans. R. Bowlby. Stanford, CA: Stanford University Press.

Derrida, Jacques. 2008. *The Animal That Therefore I Am* . Trans. David Wills. Ed. Marie-Louise Mallet. New York: Fordham University Press.

Dix, Alan. 2008. Response to "sometimes it's hard to be a robot: A call for action on the ethics of abusing artificial agents." *Interacting with Computers* 20 (3): 334–337. doi:10.1016/j.intcom.2008.02.003.

Dix, Alan. 2017. I in an other's eye. *AI & Society*. https://link.springer.com/article/10.1007/s00146-017-0694-7.

Douzinas, Costas. 2000. *The End of Human Rights: Critical Thought at the Turn of the Century*. Oxford: Hart Publishing.

Dowling, Conor M., and Michael G. Miller. 2014. *Super PAC!: Money, Elections, and Voters After Citizens United*. New York: Routledge.

Dreyfus, Hubert L. 1992. *What Computers Still Cannot Do: A Critique of Artificial Reason*. Cambridge, MA: MIT Press.

Duffy, Brian R. 2003. Anthropomorphism and the social robot. *Robotics and Autonomous Systems* 42 (3–4): 177–190. doi:10.1016/S0921-8890(02)00374-3.

Duffy, Brian R., and Gina Joue. 2004. "I, robot being," in *Intelligent Autonomous Systems Conference*. Amsterdam, Netherlands. http://citeseerx.ist.psu.edu/viewdoc/summary?doi=10.1.1.59.2833.

# References

Duncan, Roger. 2006. Emmanuel Levinas: Non-intentional consciousness and the status of representational thinking. In *Logos of Phenomenology and Phenomenology of the Logos, Book Three*, ed. Anna-Teresa Tymieniecka, 271–281. Analecta Husserliana: The Yearbook of Phenomenological Research, vol. 90. Dordrecht, the Netherlands: Springer.

Dworkin, Ronald. 1997. *Taking Rights Seriously*. Cambridge, MA: Harvard University Press.

Engineering and Physical Sciences Research Council (EPSRC). 2005. *Walking with Robots* EP/D05656X/1. http://webarchive.nationalarchives.gov.uk/20101109054809/ http://gow.epsrc.ac.uk/ViewGrant.aspx?GrantRef=EP/D05656X/1.

Epley, Nicolas, Adam Waytz, and John T. Cacioppo. 2007. On seeing human: A three-factor theory of anthropomorphism. *Psychological Review* 114 (4): 864–886. doi:10.1037/0033-295X.114.4.864.

Ess, Charles. 1996. The political computer: Democracy, cmc, and Habermas. In *Philosophical Perspectives on Computer-Mediated Communication*, ed. Charles Ess, 196–230. Albany, NY: SUNY Press.

Ess, Charles. 2009. *Digital Media Ethics*. Cambridge: Polity Press.

Ess, Charles. 2016. What's love got to do with it? Robots, sexuality, and the arts of being human. In *Social Robots: Boundaries, Potential, Challenges*, ed. Marco Nørskov, 57–79. Farmham, UK: Ashgate.

Eyssel, Friederike, and Dieta Kuchenbrandt. 2012. Social categorization of social robots: Anthropomorphism as a function of robot group membership. *British Journal of Social Psychology* 51 (4): 724–731. doi:10.1111/j.2044-8309.2011.02082.x.

Faragó, Tamás, Ádám Miklósi, Beáta Korcsok, Judit Száraz, and Márta Gácsi. 2014. Social behaviours in dog–owner interactions can serve as a model for designing social robots. *Interaction Studies: Social Behaviour and Communication in Biological and Artificial Systems* 15 (2): 143–172. doi:10.1075/is.15.2.01far.

Feenberg, Andrew. 1991. *Critical Theory of Technology*. Oxford: Oxford University Press.

Fisher, Richard. 2013. Is it OK to torture or murder a robot? *BBC Future*. http://www.bbc.com/future/story/20131127-would-you-murder-a-robot.

Floridi, Luciano. 2008. Information ethics: Its nature and scope. In *Information Technology and Moral Philosophy*, ed. Jeroen van den Hoven and John Weckert, 40–65. Cambridge: Cambridge University Press.

Floridi, Luciano. 2013. *The Ethics of Information*. Oxford: Oxford University Press.

Floridi, Luciano. 2017. Robots, jobs, taxes, and responsibilities. *Philosophy & Technology* 30 (1): 1–4. doi:10.1007/s13347-017-0257-3.

Floridi, Luciano, and J. W. Sanders. 2004. On the morality of artificial agents. *Minds and Machines* 14:349–379. doi:10.1023/B:MIND.0000035461.63578.9d.

Flynn, Clifton P. 2008. *Social Creatures: A Human and Animal Studies Reader*. New York: Lantern Books.

Ford, Martin. 2015. *Rise of the Robots: Technology and the Threat of a Jobless Future*. New York: Basic Books.

Freitas, Robert A. 1985. End note—can the wheels of justice turn for our friends in the mechanical kingdom? Don't laugh. [Also available under the title "Legal Rights of Robots" at http://www.rfreitas.com/Astro/LegalRightsOfRobots.htm.] *Student Lawyer (Chicago, Ill.)* 13 (5): 54–56.

Frey, Carl Benedikt, and Michael A. Osborne. 2017. The future of employment: How susceptible are jobs to computerisation? *Technological Forecasting and Social Change* 114:254–280. doi:10.1016/j.techfore.2016.08.019.

Garreau, Joel. 2007. Bots on the ground: In the field of battle (or even above it), robots are a soldier's best friend. *Washington Post*, 6 May 6, 2007. http://www.washingtonpost.com/wp-dyn/content/article/2007/05/05/AR2007050501009.html.

Gartner. 2013. Gartner says the Internet of Things installed base will grow to 26 billion units by 2020. http://www.gartner.com/newsroom/id/2636073.

Gaus, Gerald, and Fred D'Agostino. 2013. *The Routledge Companion to Social and Political Philosophy*. New York: Routledge.

Geraci, Robert M. 2010. *Apocalyptic AI: Visions of Heaven in Robotics, Artificial Intelligence, and Virtual Reality*. Oxford: Oxford University Press.

Gerdes, Anne. 2015. The issue of moral consideration in robot ethics. *ACM SIGCAS Computers & Society* 45 (3):274–280. doi:10.1145/2874239.2874278.

Ginway, M. Elizabeth. 2011. *Brazilian Science Fiction: Cultural Myths and Nationhood in the Land of the Future*. Lewisburg, PA: Bucknell University Press.

Gips, James. 1991. Towards the ethical robot. In *Android Epistemology*, ed. Kenneth M. Ford, Clark Glymour and Patrick Hayes, 243–252. Cambridge, MA: MIT Press.

Goertzel, Ben. 2002. Thoughts on AI morality. *Dynamical Psychology: An International, Interdisciplinary Journal of Complex Mental Processes* (May). http://www.goertzel.org/dynapsyc/2002/AIMorality.htm.

Gray, Heather M., Kurt Gray, and Daniel M. Wegner. 2007. Dimensions of mind perception. *Science* 315 (5812): 619. doi:10.1126/science.1134475.

Guizzo, Erico, and Evan Ackerman. 2016. When robots decide to kill. *IEEE Spectrum* 53 (6): 38–43. doi:10.1109/MSPEC.2016.7473151.

Gunkel, David J. 2001. *Hacking Cyberspace*. Boulder, CO: Westview Press.

# References

Gunkel, David J. 2007. *Thinking Otherwise: Philosophy, Communication, Technology.* West Lafayette, IN: Purdue University Press.

Gunkel, David J. 2010. The real problem: Avatars, metaphysics, and online social interaction. *New Media & Society* 12 (1): 127–141. doi:10.1177/1461444809341443.

Gunkel, David J. 2012. *The Machine Question: Critical Perspectives on AI, Robots, and Ethics.* Cambridge, MA: MIT Press.

Gunkel, David J. 2014a. A vindication of the rights of machines. *Philosophy & Technology* 27 (1): 113–132. doi:10.1007/s13347-013-0121-z.

Gunkel, David J. 2014b. The other question: The issue of robot rights. In *Social Robots and the Future of Social Relations*, ed. Johanna Seibt, Raul Hakli, and Marco Nørskov, 13–15. Amsterdam: IOS Press.

Gunkel, David J. 2015. The rights of machines: Caring for robotic care-givers. In *Machine Medical Ethics*, ed. Simon Peter van Rysewyk and Matthijs Pontier, 151–166. New York: Springer.

Gunkel, David J. 2016a. *Of Remixology: Ethics and Aesthetics After Remix.* Cambridge, MA: MIT Press.

Gunkel, David J. 2016b. Other problems: Rethinking ethics in the face of social robots. In *What Social Robots Can and Should Do*, ed. Johanna Seibt, Marco Nørskov, and Søren Schack Andersen, 9–12. Amsterdam: IOS Press.

Gunkel, David. J. 2016c. Another alterity: Rethinking ethics in the face of the machine. In *The Changing Face of Alterity: Communication, Technology and Other Subjects*, ed. David J. Gunkel, Ciro Marcondes Filho, and Dieter Mersch, 197–218. London: Rowman & Littlefield International.

Gunkel, David J. 2017a. The other question: Can and should robots have rights? *Ethics and Information Technology.* Published ahead of print, October 17. doi:10.1007/s10676-017-9442-4.

Gunkel, David J. 2017b. O sintoma da ética ou: Os mecanismos da filosofia moral. Trans. F. Fernandez and V. Prates. In *Sintoma e Fantasia no Capitalismo Comunicacional*, ed. Luiz Aidar Prado José and Vinicius Prates. São Paulo: Estação das Letras e Cores.

Gunkel, David J. 2018. *Gaming the System: Deconstructing Video Games, Game Studies and Virtual Worlds.* Bloomington, IN: Indiana University Press.

Gunkel, David J., and Billy Cripe. 2014. Apocalypse not, or how I learned to stop worrying and love the machine. *Kritikos* 11. http://intertheory.org/gunkel-cripe.htm.

Gunkel, David J., and Paul Taylor. 2014. *Heidegger and the Media.* New York: Polity Press.

Günther, Jan-Philipp. 2016. *Roboter und Rechtliche Verantwortung: Eine Untersuchung der Benutzer- und Herstellerhaftung*. München: Herbert Utz Verlag.

Güzeldere, Güven. 1997. The many faces of consciousness: A field guide. In *The Nature of Consciousness: Philosophical Debates*, ed. Ned Block, Owen Flanagan, and Güven Güzeldere, 1–68. Cambridge, MA: MIT Press.

Haddadin, Sami. 2014. *Towards Safe Robots: Approaching Asimov's 1st Law*. Heidelberg: Springer.

Hagendorff, Thilo. 2017. Animal rights and robot ethics. *International Journal of Technoethics* 8 (2): 61–71. doi:10.4018/IJT.2017070105.

Hajdin, Mane. 1994. *The Boundaries of Moral Discourse*. Chicago: Loyola University Press.

Hall, J. Storrs. 2001. Ethics for machines. *KurzweilAI.net*, July 5. http://www.kurzweilai.net/ethics-for-machines.

Hall, J. Storrs. 2011. *Beyond AI: Creating the Conscience of the Machine*. Amherst, NY: Prometheus Books.

Hammond, Kris. 2016. Ethics and artificial intelligence: The moral compass of a machine: The question of robotic ethics is making everyone tense. *Recode*, 13 April. https://www.recode.net/2016/4/13/11644890/ethics-and-artificial-intelligence-the-moral-compass-of-a-machine.

Hampton, Gregory Jerome. 2015. *Imagining Slaves and Robots in Literature, Film, and Popular Culture: Reinventing Yesterday's Slave with Tomorrow's Robot*. New York: Lexington Books.

Haraway, Donna J. 2008. *When Species Meet*. Minneapolis, MN: University of Minnesota Press.

Hardy, Rod, dir. 2005. *Battlestar Galactica*. Season 2, episode 5, "The Farm." Aired August 12 on the Sci-Fi Channel. http://www.imdb.com/title/tt0519788.

Harman, Graham. 2002. *Tool-Being: Heidegger and the Metaphysics of Objects*. Chicago: Open Court Press.

Hart, H. L. A. 1961. *The Concept of Law*. Oxford: Oxford University Press.

Hearne, Vicki. 2000. *Adam's Task: Calling Animals by Name*. New York: Akadine Press.

Heidegger, Martin. 1962. *Being and Time*. Trans. J. Macquarrie and E. Robinson. New York: Harper & Row.

Heidegger, Martin. 1977. *The Question Concerning Technology and Other Essays*. Trans. W. Lovitt. New York: Harper & Row.

# References

Heider, Fritz, and Marianne Simmel. 1944. An experimental study of apparent behavior. *American Journal of Psychology* 57 (2): 243–259. doi:10.2307/1416950.

Hemmersbaugh, P. A. 2016. NHTSA letter to Chris Urmson, director, self-driving car project, Google, Inc. https://isearch.nhtsa.gov/files/Google%20--%20compiled%20 response%20to%2012%20Nov%20%2015%20interp%20request%20--%204%20 Feb%2016%20final.htm.

Henderson, Mark. 2007. Human rights for robots? We're getting carried away. *The Times* (April 24). https://www.thetimes.co.uk/article/human-rights-for-robots-were-getting-carried-away-xfbdkpgwn0v.

Hernéndez-Orallo, José. 2017. *The Measure of All Minds: Evaluating Natural and Artificial Intelligence.* Cambridge: Cambridge University Press.

Heyn, E. T. 1904. Berlin's wonderful horse. *New York Times*, September 4.

Hill, Kashmir. 2014. Are child sex-robots inevitable? *Forbes* (July 14). https://www.forbes.com/sites/kashmirhill/2014/07/14/are-child-sex-robots-inevitable.

Himma, Kenneth Einar. 2009. Artificial agency, consciousness, and the criteria for moral agency: What properties must an artificial agent have to be a moral agent? *Ethics and Information Technology* 11 (1): 19–29. doi:10.1007/s10676-008-9167-5.

Hofstadter, Douglas R. 1979. *Gödel, Escher, Bach: An Eternal Golden Braid.* New York: Penguin.

Hofstadter, Douglas R. 2001. Staring emmy straight in the eye—And doing my best not to flinch. In *Virtual Music: Computer Synthesis of Musical Style*, ed. David Cope, 33–82. Cambridge, MA: MIT Press.

Hohfeld, Wesley. 1920. *Fundamental Legal Conceptions as Applied in Judicial Reasoning.* New Haven: Yale University Press.

Hubbard, F. Patrick. 2011. "Do androids dream?": Personhood and intelligent artifacts. *Temple Law Review* 83: 405–474. http://scholarcommons.sc.edu/law_facpub.

Hudson, W. Donald. 1969. *The Is/Ought Question: A Collection of Papers on the Central Problem in Moral Philosophy.* London: Macmillan.

Hume, David. 1980. *A Treatise of Human Nature.* New York: Oxford University Press.

Hymel, Alicia M., Daniel T. Levin, Jonathan Barrett, Megan Saylor, and Gautam Biswas. 2011. The interaction of children's concepts about agents and their ability to use an agent-based tutoring system. In *Human-Computer Interaction. Users and Applications*, ed. Julie A. Jacko, 580–589. New York: Springer.

Ihde, Don. 1990. *Technology and the Lifeworld: From Garden to Earth.* Bloomington, IN: Indiana University Press.

Inayatullah, Sohail, and Phil McNally. 1988. The rights of robots: Technology, culture and law in the 21st century. *Futures* 20 (2): 119–136. Also available at http://www.kurzweilai.net/the-rights-of-robots-technology-culture-and-law-in-the-21st-century.

Inayatullah, Sohail. 2001. The rights of your robots: Exclusion and inclusion in history and future. *KurzweilAI.net*. http://www.kurzweilai.net/the-rights-of-your-robots-exclusion-and-inclusion-in-history-and-future.

International Federation of Robotics (IFR). 2015. Industrial robot statistics. http://www.ifr.org/industrial-robots/statistics.

Ipsos MORI. 2006. Robo-rights: utopian dream or rise of the machines? https://web.archive.org/web/20070207230831/http://www.sigmascan.org:80/ViewIssue.aspx?IssueId=53.

ISO. 2012. ISO 8373: Robots and robotic devices—Vocabulary. https://www.iso.org/standard/55890.html.

Introna, Lucas D. 2010. The "measure of a man" and the ethos of hospitality: Towards an ethical dwelling with technology. *AI & Society* 25 (1): 93–102. doi:10.1007/s00146-009-0242-1.

Isaac, Alistair, and Will Bridewell. 2017. White lies on silver tongues: Why robots need to deceive (and how). In *Robot Ethics 2.0: From Autonomous Cars to Artificial Intelligence*, ed. Patrick Lin, Ryan Jenkins, and Keith Abney, 157–172. New York: Oxford University Press.

Jacobs, Harriet Ann. 2001. *Incidents in the Life of a Slave Girl*. New York: Dover.

Jaeger, Christopher B., and Daniel T. Levin. 2016. If asimo thinks, does roomba feel? The legal implications of attributing agency to technology. *Journal of Human-Robot Interaction* 5 (3): 3–25. doi:10.5898/JHRI.5.3.Jaeger.

James, Edward. 1990. Yellow, black, metal and tentacled: The race question in american science fiction. In *Science Fiction, Social Conflict and War*, ed. Philip J. Davies, 26–49. Manchester, UK: Manchester University Press.

James, Matt, and Kyle Scott, eds. 2008. Robots & rights: Will artificial intelligence change the meaning of human rights? *Symposium Report*. London: BioCentre. http://www.bioethics.ac.uk/cmsfiles/files/resources/biocentre_symposium_report__robots_and_rights_150108.pdf.

Jibo. 2014. https://www.jibo.com.

John, Robert. 2007. *Haftung für Künstliche Intelligenz: Rechtliche Beurteilung des Einsatzes Intelligenter Softwareagenten im E-Commerce*. Hamburg: Verlag Dr. Kovač.

Johnson, Barbara. 1981. Translator's introduction. In *Jacques Derrida, Disseminations*, vii–xxxiii. Chicago: University of Chicago Press.

Johnson, Brian David. 2011. *Science Fiction Prototyping: Designing the Future with Science Fiction.* Williston, VT: Morgan and Claypool Publishers. doi:10.2200/S00336 ED1V01Y201102CSL003.

Johnson, Deborah G. 2006. Computer systems: Moral entities but not moral agents. *Ethics and Information Technology* 8: 195–204. doi:10.1007/s10676-006-9111-5.

Jones, Raya. 2016. *Personhood and Social Robotics: A Psychological Consideration.* New York: Routledge.

Jones, Meg Leta, and Jason Millar. 2017. Hacking metaphors in the anticipatory governance of emerging technology: The case of regulating robots. In *The Oxford Handbook of Law, Regulation and Technology*, ed. Roger Brownsword, Eloise Scotford, and Karen Yeung, 573–596. Oxford: Oxford University Press.

Jordan, John. 2016. *Robots.* Cambridge, MA: MIT Press.

Kahn, Peter H., Hiroshi Ishiguro, Brian T. Gill, Takayuki Kanda, Nathan G. Freier, Rachel L. Severson, Jolina H. Ruckert, and Solace Shen. 2012. "Robovie, you'll have to go into the closet now": Children's social and moral relationships with a humanoid robot. *Developmental Psychology* 48 (2): 303–314. doi:10.1037/a0027033.

Kang, Minsoo. 2011. *Sublime Dreams of Living Machines: The Automaton in the European Imagination.* Cambridge, MA: Harvard University Press.

Kant, Immanuel. 1983. *Grounding for the Metaphysics of Morals.* Trans. J. W. Ellington. Indianapolis, IN: Hackett Publishing.

Kant, Immanuel. 1985. *Critique of Practical Reason.* Trans. L. W. Beck. New York: Macmillan.

Kaye, Lydia. 2016. Challenging sex robots and the brutal dehumanisation of women. *Campaign Against Sex Robots.* https://campaignagainstsexrobots.org/2016/02/10/challenging-sex-robots-and-the-brutal-dehumanisation-of-women.

Keynes, John M. 1963. Economic possibilities for our grandchildren. In *Essays in Persuasion*, 358–373. New York: W. W. Norton and Company.

Klein, Wilhelm E. J. 2016. Robots make ethics honest—and vice versa. *ACM SIGCAS Computers and Society* 45 (3): 261–269. doi:10.1145/2874239.2874276.

King, Edward. 2013. *Science Fiction and Digital Technologies in Argentine and Brazilian Culture.* New York: Palgrave Macmillan.

Kleeman, Jenny. 2017. The race to build the world's first sex robot. *The Guardian*, April 27. https://www.theguardian.com/technology/2017/apr/27/race-to-build-world-first-sex-robot.

Knight, Heather. 2014. How humans respond to robots: Building public policy through good design. Brookings Institute. https://www.brookings.edu/research/how-humans-respond-to-robots-building-public-policy-through-good-design.

Koops, Bert-Jaap. 2008. Criteria for normative technology: The acceptability of code as law in light of democratic and constitutional values. In *Regulating Technologies: Legal Futures, Regulatory Frames and Technological Fixes*, ed. Roger Brownsword and Karen Yeung, 157–174. Portland, OR: Hart Publishing.

Kramer, Matthew H. 1998. Rights without trimmings. In *A Debate Over Rights: Philosophical Enquiries*, ed. Matthew H. Kramer, N. E. Simmonds, and Hillel Steiner, 7–112. Oxford: Clarendon Press.

Kramer, Matthew H., N. E. Simmonds, and Hillel Steiner, ed. 1998. *A Debate Over Rights: Philosophical Enquiries*. Oxford: Clarendon Press.

Krishnan, Armin. 2016. *Killer Robots: Legality and Ethicality of Autonomous Weapons*. New York: Routledge.

Kriz, Sarah, Toni D. Damera, Pallavi Damera, and John R. Porter. 2010. Fictional robots as a data source in HRI research: Exploring the link between science fiction and interactional expectations. Paper presented at the 19th IEEE International Symposium on Robot and Human Interactive Communication, Viareggio, Italy, September 12–15, 458–63. doi:10.1109/ROMAN.2010.5598620.

Kurzgesagt–In a Nutshell. 2017. *Do Robots Deserve Rights? What if Machines Become Conscious?* https://www.youtube.com/watch?v=DHyUYg8X31c

Kurzweil, Ray. 2005. *The Singularity Is Near: When Humans Transcend Biology*. New York: Viking.

LaChat, Michel R. 1986. Artificial intelligence and ethics: An exercise in the moral imagination. *AI Magazine* 7 (2): 70–79. doi:10.1609/aimag.v7i2.540.

LaGrandeur, Kevin. 2013. *Androids and Intelligent Networks in Early Modern Literature and Culture*. New York: Routledge.

LaGrandeur, Kevin, and James J. Hughes. 2017. *Surviving the Machine Age: Intelligent Technology and the Transformation of Human Work*. Cham, Switzerland: Palgrave Macmillan.

Lalji, Nur. 2015. Can we learn about empathy from torturing robots? This MIT researcher is giving it a try. *Yes! Magazine*. http://www.yesmagazine.org/happiness/should-we-be-kind-to-robots-kate-darling.

Larriba, Ferran, Cristóbal Raya, Cecilio Angulo, Jordi Albo-Canals, Marta Díaz, and Roger Boldú. 2016. Externalising moods and psychological states in a cloud based system to enhance a pet-robot and child's interaction. *Biomedical Engineering Online* 15 (Suppl 1): 187–196. doi:10.1186/s12938-016-0180-3.

Lavender, Isiah. 2011. *Race in American Science Fiction*. Bloomington: Indiana University Press.

# References

Lee, Jason. 2017. *Sex Robots: The Future of Desire*. Cham, Switzerland: Palgrave Macmillan.

Leenes, Ronald, and Federica Lucivero. 2014. Laws on robots, laws by robots, laws in robots: Regulating robot behaviour by design. *Law, Innovation and Technology* 6 (2) 194–222. doi:10.5235/17579961.6.2.193.

Lehman-Wilzig, Sam. (1981) Frankenstein unbound: Towards a legal definition of artificial intelligence. *Futures* 13 (6): 442–457, December. doi:10.1016/0016-3287 (81)90100-2.

Leopold, Aldo. 1966. *A Sand County Almanac*. Oxford: Oxford University Press.

Leroux, Christophe, Roberto Labruto, Chiara Boscarato, Franco Caroleo, Jan-Philipp Günther, Severin Löffler, Florian Münch, et al. 2012. *Suggestion for a Green Paper on Legal Issues in Robotics*. euRobotics—The European Robotics Coordination Action. https://www.unipv-lawtech.eu/files/euRobotics-legal-issues-in-robotics-DRAFT _6j6ryjyp.pdf.

Leveringhaus, Alex. 2016. *Ethics and Autonomous Weapons*. London: Palgrave Macmillan.

Levin, Daniel T., Stephen S. Killingsworth, and Megan M. Saylor. 2008. Concepts about the capabilities of computers and robots: A test of the scope of adults' theory of mind. *Proceedings of the 3rd Annual IEEE International Workshop on Human and Robot Interaction*, 57–64. Amsterdam: ACM.

Levin, Daniel T., Stephen S. Killingsworth, Megan M. Saylor, Stephen M. Gordon, and Kazuhiko Kawamura. 2013. Tests of concepts about different kinds of minds: Predictions about the behavior of computers, robots, and people. *Human-Computer Interaction* 28 (2): 161–191.

Levinas, Emmanuel. 1969. *Totality and Infinity: An Essay on Exteriority*. Trans. A. Lingis. Pittsburgh, PA: Duquesne University.

Levinas, Emmanuel. 1981. *Otherwise Than Being, Or Beyond Essence*. Trans. A. Lingis. The Hague: Martinus Nijhoff Publishers.

Levinas, Emmanuel. 1985. *Ethics and Infinity: Conversations with Philippe Nemo*. Trans. R. A. Cohen. Pittsburgh: Duquesne University Press.

Levinas, Emmanuel. 1987. *Collected Philosophical Papers*. Trans. A. Lingis. Dordrecht: Martinus Nijhoff.

Levy, David. 2005. *Robots Unlimited: Life in a Virtual Age*. Boca Raton, FL: CRC Press.

Levy, David. 2007. *Love and Sex with Robots: The Evolution of Human-Robot Relationships*. New York: Harper Perennial.

Levy, David. 2009. The ethical treatment of artificially conscious robots. *International Journal of Social Robotics* 1 (3): 209–216. doi:10.1007/s12369-009-0022-6.

Lin, Patrick. 2016. Why ethics matters for autonomous cars. In *Autonomous Driving: Technical, Legal and Social Aspects*, ed. Markus Maurer, J. Christian Gerdes, Barbara Lenz, and Hermann Winner, 69–86. Berlin: Springer. doi:10.1007/978-3-662-48847-8_4.

Lin, Patrick, Keith Abney, and George A. Bekey. 2012. *Robot Ethics: The Ethical and Social Implications of Robotics*. Cambridge, MA: MIT Press.

Lin, Patrick, Ryan Jenkins, and Keith Abney. 2017. *Robot Ethics 2.0: New Challenges in Philosophy, Law, and Society*. New York: Oxford University Press.

Llewelyn, John. 1995. *Emmanuel Levinas: The Genealogy of Ethics*. London: Routledge.

Løgstrup, Knud Ejler. 1997. *The Ethical Demand*. Trans. H. Fink and A. MacIntyre. Notre Dame: University of Notre Dame Press.

Loughlin, Martin. 2000. *Sword and Scales: An Examination of the Relationship between Law and Politics*. Oxford: Hart Publishing.

Lovgren, Stefan. 2006. A robot in every home by 2020, South Korea says. *National Geographic News*, September 6. http://news.nationalgeographic.com/news/2006/09/060906-robots.html.

Lyotard, Jean-François. 1984. *The Postmodern Condition: A Report on Knowledge*. Trans. G. Bennington and B. Massumi. Minneapolis, MN: University of Minnesota Press.

MacCormick, Neil. 1982. *Legal Right and Social Democracy*. Oxford: Oxford University Press.

Marks, Paul. 2007. Robot rights. *New Scientist Blog* (April 24). https://www.newscientist.com/blog/technology/2007/04/robot-rights.html.

Matarić, Maja J. 2007. *The Robotics Primer*. Cambridge, MA: MIT Press.

Markus, Hazel R., and Shinobu Kitayama. 1991. Culture and the self: Implications for cognition, emotion, and motivation. *Psychological Review* 98 (2): 224–253.

Marx, Karl. 1977. *Capital: A Critique of Political Economy*. Trans. Ben Fowkes. New York: Vintage Books.

Marx, Johannes, and Christine Tiefensee. 2015. Of animals, robots and men. *Historical Social Research (Köln)* 40 (4): 70–91. doi:10.12759/hsr.40.2015.4.70-91.

McCarthy, John. 1999. Making robots conscious of their mental states. In *Machine Intelligence 15*, ed. Koichi Furukawa, Donald Michie and Stephen Muggleton, 3–17. Oxford: Oxford University.

McCauley, Lee. 2007. AI armageddon and the three laws of robotics. *Ethics and Information Technology* 9 (2): 153–164. doi:10.1007/s10676-007-9138-2.

Merriam-Webster Dictionary. 2017. Robot. https://www.merriam-webster.com/dictionary/robot.

Miller, Lantz Fleming. 2015. Granting automata human rights: Challenge to a basis of full-rights privilege. *Human Rights Review (Piscataway, N.J.)* 16 (4): 369–391. doi:10.1007/s12142-015-0387-x.

Miller, Lantz Fleming. 2017. Responsible research for the construction of maximally humanlike automata: The paradox of unattainable informed consent. *Ethics and Information Technology*. Published ahead of print, July 8. doi:10.1007/s10676-017-9427-3.

Minsky, Marvin. 2006. Alienable rights. In *Thinking about Android Epistemology*, ed. Kenneth M. Ford, Clark Glymour, and Patrick J. Hayes, 137–46. Menlo Park, CA: IAAA Press. Originally published in *Discover Magazine*, 1993.

Muehlhauser, Luke, and Nick Bostrom. 2014. Why we need friendly AI. *Think (London, England)* 13 (36): 41–47. doi:10.1017/S1477175613000316.

Nagenborg, Michael, Rafael Capurro, Jutta Weber, and Christoph Pingel. 2008. Ethical regulations on robotics in Europe. *AI & Society* 22 (3): 349–366. doi:10.1007/s00146-007-0153-y.

Nass, Clifford, Youngme Moon, John Morkes, Eun-Young Kim, and B. J. Fogg. 1997. Computers are social actors: A review of current research. In *Human Values and the Design of Computer Technology*, ed. Batya Friedman, 137–162. Cambridge: Cambridge University Press.

Nass, Clifford, and Youngme Moon. 2000. Machines and mindlessness: Social responses to computers. *Journal of Social Issues* 56 (1): 81–103. doi:10.1111/0022-4537.00153.

Nealon, Jeffrey. 1998. *Alterity Politics: Ethics and Performative Subjectivity*. Durham, NC: Duke University Press.

Neely, Erica L. 2012. Machines and the moral community. In *The Machine Question: AI, Ethics and Moral Responsibility. Proceedings of the AISB/IACAP World Congress 2012* (Birmingham, UK, July 2–6), ed. Joanna Bryson, David J. Gunkel, and Steve Torrance, 38–42. http://events.cs.bham.ac.uk/turing12/proceedings/14.pdf.

Neely, Erica L. 2014. Machines and the moral community. *Philosophy & Technology* 27 (1): 97–111. doi:10.1007/s13347-013-0114-y.

Neuhäuser, Christian. 2015. Some skeptical remarks regarding robot responsibility and a way forward. In *Collective Agency and Cooperation in Natural and Artificial Systems: Explanation, Implementation, and Simulation*, ed. Catrin Misselhorn, 131–148. New York: Springer.

Nourbakhsh, Illah. 2013. *Robot Futures*. Cambridge, MA: MIT Press.

Nørskov, Marco. 2016. *Social Robots: Boundaries, Potential, Challenges*. Farnham, UK: Ashgate.

Nyholm, Sven, and Jilles Smids. 2016. The ethics of accident-algorithms for self-driving cars: An applied trolley problem? *Ethical Theory and Moral Practice* 19 (5): 1275–1289. doi:10.1007/s10677-016-9745-2.

O'Regan, J. Kevin. 2007. How to build consciousness into a robot: The sensorimotor approach. In *50 Years of Artificial Intelligence: Essays Dedicated to the 50th Anniversary of Artificial Intelligence*. Ed. Max Lungarella, Fumiya Iida, Josh Bongard, and Rolf Pfeifer, 332–346. Berlin: Springer-Verlag.

Orwell, George. 1993. *Animal Farm*. New York: Random House/Everyman's Library.

*Oxford English Dictionary*. 2017. Robot. https://en.oxforddictionaries.com/definition/robot.

Pagallo, Ugo. 2013. *The Laws of Robots: Crimes, Contracts, and Torts*. New York: Springer.

Perkins, Franklin. 2004. *Leibniz and China: A Commerce of Light*. Cambridge: Cambridge University Press.

Pfungst. Oskar. 1965. *Clever Hans (The Horse of Mr. von Osten): A Contribution to Experimental Animal and Human Psychology*. New York: Holt, Rinehart & Winston.

Piekarski, Michał. 2016. The problem of the question about animal ethics: Discussion with Mark Coeckelbergh and David Gunkel. *Journal of Agricultural & Environmental Ethics* 29 (4): 705–715. doi:10.1007/s10806-016-9626-7.

Plato. 1977. *Plato IV: Cratylus, Parmenides, Greater Hippias, Lesser Hippias*. Trans. H. N. Fowler. Cambridge, MA: Harvard University Press.

Plato. 1981. *Plato IX: Timaeus, Critias, Cleitophon, Menexenus, Epistles*. Trans. R. G. Bury. Cambridge, MA: Harvard University Press.

Plato. 1982. *Plato I: Euthyphro, Apology, Crito, Phaedo, Phaedrus*. Trans. H. N. Fowler. Cambridge, MA: Harvard University Press.

Plato. 1987. *Plato V & VI: Republic*. Trans. P. Shorey. Cambridge, MA: Harvard University Press.

Prescott, Tony J. 2017. Robots are not just tools. *Connection Science* 29 (2): 142–149. doi:10.1080/09540091.2017.1279125.

Prescott, Tony J., and Michael Szollosy. 2017. Ethical principles of robotics. *Connection Science* 29 (2): 119–123. doi:10.1080/09540091.2017.1312800.

Putnam, Hilary. 1964. Robots: Machines or artificially created life? *Journal of Philosophy* 61 (21): 668–691. http://www.jstor.org/stable/2023045.

Randerson, James. 2007. Forget robot rights, experts say, use them for public safety. *The Guardian*, April 23. https://www.theguardian.com/science/2007/apr/24/frontpagenews.uknews.

Raz, Joseph. 1986. *The Morality of Freedom*. Oxford: Oxford University Press.

Raz, Joseph. 1994. *Ethics in the Public Domain*. Oxford: Oxford University Press.

Read, Brock. 2007. The wrong take on "robot rights"? *The Chronicle of Higher Education*. http://www.chronicle.com/blogs/wiredcampus/the-wrong-take-on-robot-rights/2984.

Regan, Tom. 1983. *The Case for Animal Rights*. Berkeley, CA: University of California Press.

Reeves, Byron, and Clifford Nass. 1996. *The Media Equation: How People Treat Computers, Television, and New Media Like Real People and Places*. Cambridge: Cambridge University Press.

Richards, Neil M., and William D. Smart. 2017. How should the law think about robots? In *Robot Law*, ed. Ryan Calo, A. Michael Froomkin, and Ian Kerr, 3–24. Cheltenham, UK: Edward Elgar.

Richardson, Kathleen. 2015. The asymmetrical "relationship": Parallels between prostitution and the development of sex robots. *SIGCAS Computers & Society* 45 (3): 290–293. https://campaignagainstsexrobots.org/the-asymmetrical-relationship-parallels-between-prostitution-and-the-development-of-sex-robots.

Richardson, Kathleen. 2016a. Sex robot matters: Slavery, the prostituted, and the rights of machines. *IEEE Technology and Society Magazine* 35 (2): 46–53. doi:10.1109/MTS.2016.2554421.

Richardson, Kathleen. 2016b. Are sex robots as bad as killing robots? In *What Social Robots Can and Should Do*, ed. Johanna Seibt, Marco Nørskov, and Søren Schack Andersen, 27–31. Amsterdam: ISO Press.

Richardson, Kathleen. 2017. Rethinking the I-you relation through dialogical philosophy in the ethics of AI and robotics. *AI & Society*. Published ahead of print, March 9. doi:10.1007/s00146-017-0703-x.

Ricœur, Paul. 2007. *Reflections on the Just*. Trans. D. Pellauer. Chicago: University of Chicago Press.

Riek, Laurel D., Tal-Chen Rabinowitch, Bhismadev Chakrabarti, and Peter Robinson. 2009. How anthropomorphism affects empathy toward robots. *HRI '09 Proceedings of the 4th ACM/IEEE International Conference on Human Robot Interaction* (La Jolla, CA, March 9–13), 245–246. New York: ACM. doi:10.1145/1514095.1514158.

Robertson, Jennifer. 2007. Robo sapiens japanicus: Humanoid robots and the posthuman family. *Critical Asian Studies* 39 (3): 369–398. doi:10.1080/14672710701527378.

Robertson, Jennifer. 2014. Human rights vs. robot rights: Forecasts from japan. *Critical Asian Studies* 46 (4): 571–598. doi:10.1080/14672715.2014.960707.

Robertson, Jennifer. 2017. *Robo Sapiens Japanicus: Robots, Gender, Family, and the Japanese Nation.* Berkeley: University of California Press.

Roh, Daniel. 2009. Do humanlike machines deserve human rights? *Wired.* https://www.wired.com/2009/01/st-essay-16.

Rosenthal-von der Pütten, Astrid M., Nicole C. Krämer, Laura Hoffmann, Sabrina Sobieraj, and Sabrina C. Eimler. 2013. An experimental study on emotional reactions towards a robot. *International Journal of Social Robotics* 5: 17–34. doi:10.1007/s12369-012-0173-8.

Ross, P. E. 2016. A Google car can qualify as a legal driver. *IEEE Spectrum.* http://spectrum.ieee.org/cars-that-think/transportation/self-driving/an-ai-can-legally-be-defined-as-a-cars-driver.

Rubin, Charles T. 2011. Machine morality and human responsibility. *New Atlantis (Washington, D.C.)* 32:58–79. http://www.jstor.org/stable/43152657.

Said, Edward W. 1978. *Orientalism.* New York: Random House.

Samani, Hooman Aghaebrahimi, Adrian David Cheok, Mili John Tharakan, Jeffrey Koh, and Newton Fernando. 2010. A design process for lovotics. In *Human-Robot Personal Relationships*, ed. Maarten H. Lamers and Fons J. Verbeek, 118–125. Heidelberg: Springer.

Sandoval, Eduardo Benitez, Omar Mubin, and Mohammad Obaid. 2014. Human robot interaction and fiction: A contradiction. In *Social Robotics: 6th International Conference, ICSR 2014,* Sydney, Australia, 27–29 October, ed. Michael Beetz, Benjamin Johnston, and Mary-Anne Williams, 54–63. Cham, Switzerland: Springer.

Sandry, Eleanor. 2015a. Re-evaluating the form and communication of social robots: The benefits of collaborating with machinelike robots. *International Journal of Social Robotics* 7 (3): 335–346. doi:10.1007/s12369-014-0278-3.

Sandry, Eleanor. 2015b. *Robots and Communication.* New York: Palgrave Macmillan.

Santosuosso, Amedeo. 2016. The human rights of nonhuman artificial entities: An oxymoron? *Jahrbuch für Wissenschaft und Ethik* 19 (1): 203–238. doi:10.1515/jwiet-2015-0114.

de Saussure, Ferdinand. 1959. *Course in General Linguistics.* Trans. W. Baskin. London: Peter Owen.

Scheutz, Matthias. 2012. The inherent dangers of unidirectional emotional bonds between humans and social robots. In *Robot Ethics: The Ethical and Social Implications of Robotics*, ed. Patrick Lin, Keith Abney, and George A. Bekey, 205–221. Cambridge, MA: MIT Press.

Schurz, Gerhard. 1997. *The Is-Ought Problem: An Investigation in Philosophical Logic*. Dordrecht: Springer.

Schweighofer, Erich. 2001. Vorüberlegungen zu künstlichen personen: Autonome roboter und intelligente softwareagenten. In *Auf dem Weg zur ePerson: Aktuelle Fragestellungen der Rechtsinformatik, Nr. 3 der Reihne Schriftenreihe Rechtsinformatik*, ed. Erich Scheighofer, Thomas Menzel, and Günther Kreuzbauer, 45–53. Wien: Verlag Österreich.

Schwitzgebel, Eric, and Mara Garza. 2015. A defense of the rights of artificial intelligences. *Midwest Studies in Philosophy* 39 (1): 89–119. doi:10.1111/misp.12032.

Scott, Robert L. 1967. On viewing rhetoric as epistemic. *Central States Speech Journal* 18: 9–17. doi:10.1080/10510976709362856.

Scott, Robert L. 1976. On viewing rhetoric as epistemic: Ten years later. *Central States Speech Journal* 27 (4): 258–266. doi:10.1080/10510977609367902.

Searle, John R. 1964. How to derive "ought" from "is." *Philosophical Review* 73 (1): 43–58. http://www.jstor.org/stable/2183201.

Searle, John R. 1980. Minds, brains and programs. *Behavioral and Brain Sciences* 3 (3): 417–57. doi:10.1017/S0140525X00005756.

Searle, John. 1997. *The Mystery of Consciousness*. New York: New York Review of Books.

Searle, John. 1999. The Chinese room. In *The MIT Encyclopedia of the Cognitive Sciences*. eds. R. A. Wilson,, and F. Keil, 115–116. Cambridge, MA: MIT Press.

Seibt, Johanna. 2017. Towards an ontology of simulated social interaction: Varieties of the "as if" for robots and humans. In *Sociality and Normativity for Robots—Philosophical Investigations* , ed. Raul Hakli and Johanna Seibt, 11–39. Studies in the Philosophy of Sociality. Cham, Switzerland: Springer. doi:10.1007/978-3-319-53133-5.

Seibt, Johanna, Raul Hakli, and Marco Nørskov. 2014. *Sociable Robots and the Future of Social Relations: Proceedings of Robophilosophy 2014*. Amsterdam: IOS Press.

Seibt, Johanna, Marco Nørskov, and Søren Schack Andersen. 2016. *What Social Robots Can and Should Do: Proceedings of Robophilosophy 2016*. Amsterdam: IOS Press.

Sharkey, Noel E. 2012. The inevitability of autonomous robot warfare. *International Review of the Red Cross* 94 (886): 787–799. doi:10.1017/S1816383112000732.

Sharkey, Amanda, and Noel E. Sharkey. 2010. The crying shame of robot nannies: An ethical appraisal. *Interaction Studies: Social Behaviour and Communication in Biological and Artificial Systems* 11 (2): 161–190. doi:10.1075/is.11.2.01sha.

Sharkey, Amanda, and Noel E. Sharkey. 2012. Granny and the robots: Ethical issues in robot care for the elderly. *Ethics and Information Technology* 14 (1): 27–40. doi:10.1007/s10676-010-9234-6.

Sharkey, Noel E., and Eduard Fosch-Villaronga. 2017. FRR highlights new European commission report on robot law. *Foundation for Responsible Robotics*, 20 February. http://responsiblerobotics.org/frr-highlights-ec/f.

Sharkey, Noel E., Aime van Wynsberghe, Scott Robbins, and Eleanor Hancock. 2017. Our sexual future with robots: A foundation for responsible robotics consultation report. *Foundation for Responsible Robotics*. https://responsible-robotics-myxf6pn3xr.netdna-ssl.com/wp-content/uploads/2017/11/FRR-Consultation-Report-Our-Sexual-Future-with-robots-1-1.pdf.

Sherry, John L. 2001. The effect of violent video games on aggression: A meta-analysis. *Human Communication Research* 27 (3): 409–431. doi:10.1111/j.1468-2958.2001.tb00787.x.

Sherry, John L. 2006. Would the great and mighty Oz play doom? A Look behind the curtain of violent video game research. In *Digital Media: Transformations in Human Communication*, ed. Paul Messaris and Lee Humphreys, 225–236. New York: Peter Lang.

Shneiderman, Ben. 1989. A nonanthropomorphic style guide: Overcoming the humpty–dumpty syndrome. *The Computing Teacher* 16 (7): 5–7.

Simon, Matt. 2017. What is a robot? *Wired*, August 24. https://www.wired.com/story/what-is-a-robot.

Singer, Peter Warren. 2009. *Wired for War: The Robotics Revolution and Conflict in the Twenty-First Century*. New York: Penguin Books.

Singer, Peter. 1975. *Animal Liberation: A New Ethics for Our Treatment of Animals*. New York: New York Review of Books.

Singer, Peter. 1989. All animals are equal. In *Animal Rights and Human Obligations*, ed. Tom Regan and Peter Singer, 148–162. Upper Saddle River, NJ: Prentice-Hall.

Singer, Peter, and Agata Sagan. 2009. When robots have feelings. *The Guardian*, December 14. https://www.theguardian.com/commentisfree/2009/dec/14/rage-against-machines-robots.

Siracusa, John, and Jason Snell. 2015. Robot or not? https://www.theincomparable.com/robot.

Sloman, Aaron. 2010. Requirements for artificial companions: It's harder than you think. In *Close Engagements with Artificial Companions: Key Social, Psychological, Ethical, and Design Issues*, ed. Yorick Wilks, 179–200. Amsterdam: John Benjamins.

Smith, Wesley J. 2015. AI machine: Things not persons. *First Things*. https://www.firstthings.com/web-exclusives/2015/04/ai-machines-things-not-persons.

Solaiman, S. M. 2017. Legal personality of robots, corporations, idols and chimpanzees: A quest for legitimacy. *Artificial Intelligence and Law* 25 (2): 155–179. doi:10.1007/s10506-016-9192-3.

# References

Solum, Lawrence B. 1992. Legal personhood for artificial intelligences. *North Carolina Law Review* 70 (4): 1231–1287. http://scholarship.law.unc.edu/nclr/vol70/iss4/4.

Sparrow, Robert. 2004. The turing triage test. *Ethics and Information Technology* 6 (4): 203–213. doi:10.1007/s10676-004-6491-2.

Sparrow, Robert. 2007. Killer robots. *Journal of Applied Philosophy* 24 (1): 62–77. doi:10.1111/j.1468-5930.2007.00346.x.

Sparrow, Robert. 2012. Can machines be people? Reflections on the turing triage test. In *Robot Ethics: The Ethical and Social Implications of Robotics*, ed. Patrick Lin, Keith Abney and George A. Bekey, 301–316. Cambridge, MA: MIT Press.

Sparrow, Robert, and Linda Sparrow. 2006. In the hands of machines? The future of aged care. *Mind and Machine* 16 (2): 141–161. doi:10.1007/s11023-006-9030-6.

Stanley. 2015. *Matter of Nonhuman Rights Project, Inc v Stanley*. NY Slip Op 25257 [49 Misc 3d 746]. http://www.courts.state.ny.us/REPORTER/3dseries/2015/2015_25257.htm.

Steiner, Hillel. 1994. *An Essay on Rights*. Oxford: Blackwell.

Steiner, Hillel. 1998. Working rights. In *A Debate Over Rights: Philosophical Enquiries*, ed. Matthew H. Kramer, N. E. Simmonds, and Hillel Steiner, 113–232. Oxford: Clarendon Press.

Steiner, Hillel. 2006. Moral rights. In *The Oxford Handbook of Ethical Theory*, ed. David Copp, 459–479. Oxford: Oxford University Press.

Stone, Christopher D. 1974. *Should Trees Have Standing? Toward Legal Rights for Natural Objects*. Los Altos, CA: William Kaufmann.

Stork, David G., ed. 1997. *Hal's Legacy: 2001's Computer as Dream and Reality*. Cambridge, MA: MIT Press.

Sudia, Frank W. 2001. A jurisprudence of artilects: Blueprint for a synthetic citizen. *Journal of Future Studies* 6 (2): 65–80. Also available at http://jfsdigital.org/wp-content/uploads/2014/06/062-A04.pdf and http://www.kurzweilai.net/a-jurisprudence-of-artilects-blueprint-for-a-synthetic-citizen.

Sullins, John P. 2005. The ambiguous ethical status of autonomous robots. American Association for Artificial Intelligence. Fall Symposia, Arlington, Virginia (4–6 November). https://www.aaai.org/Papers/Symposia/Fall/2005/FS-05-06/FS05-06-019.pdf.

Sullins, John P. 2012. Robots, love, and sex: The ethics of building a love machine. *IEEE Transactions on Affective Computing* 3 (4): 398–409. doi:10.1109/T-AFFC.2012.31.

Sumner, L. W. 1987. *The Moral Foundation of Rights*. Oxford: Charenden Press.

Sumner, L. W. 2000. Rights. In *The Blackwell Guide to Ethical Theory*, ed. Hugh LaFolletee, 288–305. Oxford: Blackwell Publishers.

Sung, Ja-Young, Lan Guo, Rebecca E. Grinter, and Henrik I. Christensen. 2007. "My roomba is rambo": Intimate home appliances. In *Proceedings of UbiComp 2007*, ed. John Krumm, Gregory D. Abowd, Aruna Seneviratne, and Thomas Strange, 145–162. Berlin: Springer-Verlag.

Sunstein, Cass R. 2003. The rights of animals. *University of Chicago Law Review* 70 (1): 387–401. http://www.jstor.org/stable/1600565.

Suzuki, Yutaka, Lisa Galli, Ayaka Ikeda, Shoji Itakura, and Michiteru Kitazaki. 2015. Measuring empathy for human and robot hand pain using electroencephalography. *Scientific Reports* 5: 15924. http://www.nature.com/articles/srep15924.

Szollosy, Michael. 2017. EPSRC principles of robotics: Defending an obsolete human(ism)? *Connection Science* 29 (2): 150–159. doi:10.1080/09540091.2017.1279126.

Taylor, Mark. 1997. *Hiding*. Chicago: University of Chicago Press.

Taylor, Mark. 1999. *About Religion: Economies of Faith in Virtual Culture*. Chicago: University of Chicago Press.

Teubner, Gunther. 2006. Rights of non-humans? Electronic agents and animals as new actors in politics and law. *Journal of Law and Society* 33 (4): 497–521. doi:10.1111/j.1467-6478.2006.00368.x.

Thimbleby, Harold. 2008. Robot ethics? Not yet: A reflection on Whitby's "sometimes it's hard to be a robot." *Interacting with Computers* 20 (3): 338–341. doi:10.1016/j.intcom.2008.02.006.

de Tocqueville, Alexis. 1899. *Democracy in America*. Trans. H. Reeve. New York: The Colonial Press.

Torrance, Steve. 2003. Could we, should we, create conscious robots? *Journal of Health, Social and Environmental Issues* 4 (2): 43–46. https://eprints.mdx.ac.uk/3001/1/vol4_no2.pdf.

Torrance, Steve. 2008. Ethics and consciousness in artificial agents. *AI & Society* 22 (4): 495–521. doi:10.1007/s00146-007-0091-8.

Turkle, Sherry. 1995. *Life on the Screen: Identity in the Age of the Internet*. New York: Simon and Schuster.

Turkle, Sherry. 2011. *Alone Together: Why We Expect More from Technology and Less from Each Other*. New York: Basic Books.

Turkle, Sherry, Cynthia Breazeal, Olivia Dasté, and Brian Scassellati. 2006. First encounters with kismet and cog. In *Digital Media: Transformations in Human Communication*, ed. Paul Messaris and Lee Humphreys, 313–330. New York: Peter Lang.

Turing, Alan M. 1999. Computing machinery and intelligence. In *Computer Media and Communication*, ed. Paul A. Mayer, 37–58. Oxford: Oxford University Press.

Vallor, Shannon. 2015. Moral deskilling and upskilling in a new machine age: Reflections on the ambiguous future of character. *Philosophy & Technology* 28 (1): 107–124. doi:10.1007/s13347-014-0156-9.

Velmans, Max. 2000. *Understanding Consciousness*. London: Routledge.

Versenyi, Laszlo. 1974. Can robots be moral? *Ethics* 84 (3): 248–259. http://www.jstor.org/stable/2379958.

Veruggio, Gianmarco. 2005. The birth of roboethics. IEEE International Conference on Robotics and Automation. Workshop on Robo-Ethics. Barcelona, Spain, April 18. http://www.roboethics.org/icra2005/veruggio.pdf.

Veruggio, Gianmarco. 2006. *EURON Roboethics Roadmap*. http://www.roboethics.org/atelier2006/docs/ROBOETHICS%20ROADMAP%20Rel2.1.1.pdf.

Veruggio, Gianmarco, and Fiorella Operto. 2008. Roboethics: Social and ethical implications of robotics. In *Springer Handbook of Robotics*, ed. Bruno Siciliano and Oussama Khatib, 1499–1524. New York: Springer.

Vize, Brendan. 2011. Do androids dream of electric shocks? Utilitarian machine. Master of Arts Thesis, Victoria University of Wellington. http://researcharchive.vuw.ac.nz/bitstream/handle/10063/1686/thesis.pdf?sequence=2.

Vladeck, David C. 2014. Machines without principals: Liability rules and artificial intelligence. *Washington Law Review* 89 (1): 117–150. http://hdl.handle.net/1773.1/1322.

Wallach, Wendell, and Colin Allen. 2009. *Moral Machines: Teaching Robots Right from Wrong*. Oxford: Oxford University Press.

Warwick, Kevin. 2012. Robots with biological brains. In *Robot Ethics: The Ethical and Social Implications of Robotics*, ed. Patrick Lin, Keith Abney and George A. Bekey, 317–332. Cambridge, MA: MIT Press.

Wenar, Leif. 2005. The nature of rights. *Philosophy & Public Affairs* 33 (3): 223–252. http://www.jstor.org/stable/3557929.

Wenar, Leif. 2015. Rights. *The Stanford Encyclopedia of Philosophy*, ed. Edward N. Zalta. https://plato.stanford.edu/archives/fall2015/entries/rights.

Weng, Yueh-Hsuan, Chien-Hsun Chen, and Chuen-Tsai Sun. 2009. Toward the human—robot co-existence society: On safety intelligence for next generation robots. *International Journal of Social Robotics* 1:267–282. doi:10.1007/s12369-009-0019-1.

Whitby, Blay. 2008. Sometimes it's hard to be a robot: A call for action on the ethics of abusing artificial agents. *Interacting with Computers* 20 (3): 326–333. doi:10.1016/j.intcom.2008.02.002.

Whitby, Blay. 2010. Oversold, unregulated, and unethical: Why we need to respond to robot nannies. *Interaction Studies: Social Behaviour and Communication in Biological and Artificial Systems* 11 (2): 290–294. doi:10.1075/is.11.2.18whi.

Whitby, Blay. 2011. Do you want a robot lover? The ethics of caring technologies. In *Robot Ethics: The Ethical and Social Implications of Robotics*, ed. Patrick Lin, Keith Abney and George A. Bekey, 233–248. Cambridge, MA: MIT Press.

White, Trevor N., and Seth D. Baum. 2017. Liability for present and future robotic technology. In *Robot Ethics 2.0: New Challenges in Philosophy, Law, and Society*, ed. Patrick Lin, Ryan Jenkins, and Keith Abney, 66–79. New York: Oxford University Press.

Wiener, Norbert. 1954. *The Human Use of Human Beings: Cybernetics and Society*. New York: Da Capo Press.

Wiener, Norbert. 1996. *Cybernetics: Or Control and Communication in the Animal and the Machine*. Cambridge, MA: MIT Press.

Winfield, Alan. 2007. The rights of robots. *Alan Winfield's Web Log*. http://alanwinfield.blogspot.com/2007/02/rights-of-robot.html.

Winfield, Alan. 2011a. Roboethics—For humans. *New Scientist* 210 (2811): 32–33. doi:10.1016/S0262-4079(11)61052-X.

Winfield, Alan. 2011b. An interview: Alan Winfield. *Science Museum Blog*, 30 November. https://blog.sciencemuseum.org.uk/in-interview-alan-winfield.

Winfield, Alan. 2012. *Robotics: A Very Short Introduction*. Oxford: Oxford University Press.

van Wynsberghe, Aimee. 2016. *Healthcare Robots: Ethics, Design, and Implementation*. New York: Routledge.

Yampolskiy, Roman V. 2013. Artificial intelligence safety engineering: Why machine ethics is a wrong approach. In *Philosophy and Theory of Artificial Intelligence*, ed. Vincent C. Müller, 389–396. Berlin: Springer-Verlag.

Yampolskiy, Roman V. 2016. *Artificial Superintelligence: A Futuristic Approach*. London: CRC Press.

Yonck, Richard. 2017. *Heart of the Machine: Our Future in a World of Artificial Emotional Intelligence*. New York: Arcade Publishing.

Yudkowsky, Eliezer. 2001. *Creating Friendly AI 1.0: The Analysis and Design of Benevolent Goal Architectures*. The Singularity Institute. San Francisco, CA. June 15.

Zeifman, Igal. 2017. Bot traffic report 2016. *Incapsula*. https://www.incapsula.com/blog/bot-traffic-report-2016.html.

Žižek, Slavoj. 2003. *The Puppet and the Dwarf: The Perverse Core of Christianity*. Cambridge, MA: MIT Press.

Žižek, Slavoj. 2006a. *The Parallax View*. Cambridge, MA: MIT Press.

Žižek, Slavoj. 2006b. Philosophy, the "unknown knowns," and the public use of reason. *Topoi* 25 (1–2): 137–142. doi:10.1007/s11245-006-0021-2.

Žižek, Slavoj. 2008a. *For They Know Not What They Do: Enjoyment as a Political Factor*. London: Verso.

Žižek, Slavoj. 2008b. *In Defense of Lost Causes*. London: Verso.

# Index

Anthropocentrism, 79, 177–179, 184
Anthropomorphism, 73–74, 133, 136, 145–150, 175
Appearance/reality distinction, 57–58, 145–146, 149. *See also* Plato
Aristotle, 69, 118, 120, 196 n7
Artificial intelligence (AI), 81–82; 89, 169. *See also* Robot
strong/weak AI (*see* Searle, John)
Artificial moral agent (AMA), 2, 39–40, 79–81, 107, 146–147, 171, 193n5
Asceticism, 111, 194–194n3
Asimov, Isaac, 117–118

*Battlestar Galactica*, 60, 192n3
Being vs. appearance. *See* Appearance/reality distinction
Bekey, George, 20–21, 24
Bryson, Joanna, 107–119, 123, 126, 142–143, 194n2, 196n7, 197–198n11. *See also* Robot-as-slave model

Can/should distinction. *See* Is/ought
Čapek, Karel, 1, 55–56. *See also Rossum's Universal Robots (R.U.R.)*
Consciousness
definitions of, 98–101
of machines/robots, 36, 84–85, 94, 101–102, 126
and rights/moral status, 88–90, 98, 154
self-, 91

Darling, Kate
and human robot interactions, 133–136, 140, 145, 150–152 (*see also* Human-robot interaction [HRI])
and robots as tools, 72
and robot rights, 133, 136–140, 142, 147, 156–158, 175
studies/workshops, 134–136, 143–145
Deconstruction, 8–10. *See also* Derrida, Jacques
double science, 9–10
Dennett, Daniel, 99–100, 146, 185
Derrida, Jacques, 48, 59. *See also* Poststructuralism
deconstruction, 8–10; 59
ethics, 165, 190n11
the "other," 171–172, 176, 178
de Saussure, Ferdinand, 161–162

Ethics/ethical; 50–51, 59–60, 163–166, 174–177, 183–185. *See also* Levinas, Emmanuel; Machine ethics (ME); Moral/morality; Roboethics
of artificial intelligence (AI), 111–112
normative, 113
pluralism, 181–182
relativism, 75, 180–181, 183
Ethnocentrism, 75, 115–117
Engineering and Physical Science Research Council (EPSRC), 62, 65, 117. *See also* Principles for Robotics

EURON *RoboEthics Roadmap*, 40–41. *See also* Roboethics

Floridi, Luciano, 37–38, 61
  on ethics, 2, 59, 140–141, 159, 163, 181
  on the legal status of robots, 119–123
Foundation for Responsible Robotics, 37

Heidegger, Martin, 8, 14; 53–54; 70, 191–192n2
Hegel, G. W. F., 70, 127
Hohfeld, Wesley, 26–27, 188–189n2
Hohfeldian incidents, 4, 27–30, 49, 176
Human-robot interaction (HRI), 72–74, 114–115, 129, 150–154, 174, 198n2. *See also* Darling, Kate; Sex robot/machine
Hume, David, 2–4. *See also* Is/ought
  Humean thesis, 5–6, 10, 136, 145
  Hume's Guillotine, 2–3, 10
  *A Treatise of Human Nature*, 1

Instrumentalist theory (of technology), 54–55, 69–72; 76–77, 108, 116, 172, 182
Interest Theory (of rights), 46, 49, 134. *See also* Will Theory (of rights)
Is/ought, 130, 136, 145, 159, 166, 175. *See also* Hume, David
  derivations, 10, 69, 88
  inference/problem, 2–5, 8, 65, 95
  *is* from *ought*, 8, 107
  *ought* from *is*, 6, 8, 65–68, 76, 81–88, 95, 106

Jibo, 139, 171–174. *See also* Social robots
Jordan, John, 14–16, 21–22

Kant, Immanuel, 97, 142, 150–151

Levinas, Emmanuel, 159–161, 163–168, 170–180, 184–185, 201–202n6

Levy, David, 84–85, 101–103, 126–127, 153–154
Lyotard, Jean-François, 54, 172

Machine ethics, 38, 43–44, 70, 109
Marx, Karl, 70–71
Maximally humanlike automata (MHA), 66–67, 104–105, 108
Media effect, 135
Miller, Lantz Fleming, 66–69, 104–105, 108, 110–116, 128, 194n2
Moral/morality. *See also* Artificial moral agent (AMA); Hohfeldian Incidents; Levinas, Emmanuel
  agent/patient, 2, 146–147, 177
  and robots, 41, 43–44, 50
  status, 69, 96–98, 100, 103–104, 165, 170
  systems, 3–5, 51, 75, 180–181, 183

Other (Otherness), 151, 161, 164, 166–168, 171–180 *passim*, 183–184. *See also* Levinas, Emmanuel
  the excluded/marginalized; 42, 51, 59, 103

Parallax view. *See* Žižek, Slavoj
Plato, 106
  The Allegory of the Cave; 57–58
  *The Apology*, 6–7
  *Timaeus*, 189 n7
Poststructuralism, 161–162, 164. *See also* Derrida, Jacques
Prescott, Tony, 65–66, 72–74, 172–174, 193n6
Principles for Robotics, 62–66; 68, 75, 113, 117, 193n6. *See also* EPSRC
Putnam, Hilary, 79–80, 82

Rights. *See also* Hofeldian Incidents; Interest Theory (of rights); Will Theory (of rights)
  of animals, 42, 97, 151
  definition of, 13, 26–30, 176

of human users, 48, 67–68, 176
of robots (*see* Robot rights)
Roboethics, 33–46, 84–85, 117, 152, 160–161
Robot. *See also* Artificial Intelligence (AI); Robot-as-slave model; Roboethics
　abuse, 45–46, 61, 119, 133–137, 151–157, 198n2, 198–199n3
　definitions of, 13–16, 19–26; 47; 51
　etymology of, 15, 33
　intelligence (*see* Artificial Intelligence [AI])
　law, 48, 107–108
　legal status of, 46–47, 64, 90–91, 111
　moral standing/status of, 83–88, 92–93, 104, 110, 170
　rights, *(see* Robot rights)
　sense-think-act paradigm, 20–21
　as tools, 70–73, 109–110, 154
Robot-as-slave model, 107–109, 115, 120–121, 124–127, 130–131. *See also* Bryson, Joanna; Miller, Lantz Fleming; Slavery (2.0);
Robot rights, 33–40; 43–51; 60–61, 80, 105–109, 152–153, 184–185
*Rossum's Universal Robots (R.U.R.)*, 15, 33, 55–57, 59–62. *See also* Čapek, Karel

Science fiction, 15–20, 118–120, 188n1, 196–197n8. *See also* Battlestar Galactica, *Rossum's Universal Robots (R.U.R.)*
Searle, John
　Chinese Room thought experiment, 58, 102, 150
　strong/weak artificial intelligence, 12, 188n6
Service robots, x, 9
Sex robots/machines, 128–129, 156, 197n9, 197–198n11, 199n6. *See also* Human-robot interaction (HRI)

Singer, Peter, 13, 60, 93–94, 97
Slavery (2.0), 117–125, 127–131. *See also* Bryson, Joanna; Miller, Lantz Fleming; Robot-as-slave model
Social robots, x, 114, 133, 136–139, 149, 173–175, 199n6. *See also* Human-robot interactions (HRI); Jibo
Socrates. *See Plato*

Technology. *See also* Instrumentalist theory (of technology); Robot
　applied to philosophy, 108, 160–161, 170
　definitions of, 54–55, 64, 74, 76
　effects of, 74, 144, 172
　and ethics, 38, 45, 120
　and robots, 16, 21, 23, 65, 93–94
Thinking otherwise, 8, 159, 164
Turing Test, 85, 145, 169–170
Turkle, Sherry, 72, 129, 144, 199n4

Who/what distinction, 50, 59–60, 98, 139–143, 167–169, 172–174
Will Theory (of rights), 37, 46. *See also* Interest Theory (of rights)

Žižek, Slavoj, 33, 168, 170, 180–181, 185–186
　parallax view, 33, 50